Advances in Nanomaterials

Ganesh Balasubramanian
Editor

Advances in Nanomaterials

Fundamentals, Properties and Applications

Springer

Editor
Ganesh Balasubramanian
Lehigh University
Bethlehem, Pennsylvania, USA

ISBN 978-3-319-87853-9 ISBN 978-3-319-64717-3 (eBook)
DOI 10.1007/978-3-319-64717-3

Softcover reprint of the hardcover 1st edition 2017

Printed on acid-free paper

This Springer imprint is published by Springer Nature
The registered company is Springer International Publishing AG
The registered company address is: Gewerbestrasse 11, 6330 Cham, Switzerland

Preface

Let's imagine we have a block of gold and a pair of metal-cutting scissors, and we start cutting the gold into small pieces. The original block of gold is typically of millimeter dimension, i.e., macro or bulk material. If we keep on cutting the gold to smaller and smaller sizes, such that the smallest piece is 1 millionth in size of the original block, that dimension is a nanometer, and now that "nano" piece of gold is a nanomaterial. The properties of a nanomaterial are significantly different from its bulk counterpart. For example, bulk gold melts at above 1000 °C while a gold nanoparticle melts at around room temperature. Also, gold as we know it is yellowish in color, but a nano-piece of gold is typically ruby red in color due to different light absorption properties. The fundamental discovery that material properties are different, and in many cases, better when nanostructured rather than as macroscopic particles has paved the way for a widespread interest in nanoscience.

Just to give a perspective, the thickness of an average human hair is about 100,000 nm while the diameter of an atom is about 0.4 nm. Thus, nanotechnology, in essence, is really looking at the physical and chemical mechanisms of atoms and molecules, and more importantly, as presented in this book, strategies to control and engineer these properties to create new and advanced materials that can improve our lives. The 2010 Nobel Prize in Physics for the 2004 discovery of a two-dimensional nanomaterial of carbon called graphene with remarkable electrical, thermal, and mechanical properties catapulted the interest and scientific funding in nanomaterials, not only in the developed countries, but across the globe. Interestingly, several layers of graphene when put together make graphite, which is what pencils, that we so regularly use to write, contain.

The history of nanoscience dates back to 1959, when Richard Feynman gave his extensively referred talk on molecular machines with atomic precision. It was two decades later when Eric Drexler presented molecular nanotechnology concepts at MIT. Since then, research in nanomaterials has consistently found academic interest with the discovery of buckyball in 1985 by Smalley and his colleagues and the 1991 discovery of carbon nanotubes by Ijima in Japan. Research in nanotechnology has grown exponentially over the last two decades, especially with significant commercial interest in products containing nanoparticles. For example, several of our everyday products contain nanomaterials—many sunscreens contain zinc oxide and titanium oxide nanoparticles, nano-silver contained in bandages and dressings used to treat wounds and burns, nanoparticle suspensions in fluids used as coolants for cars—as well as for certain ambitious applications such as lightweight carbon nanotube containing composite materials for aerospace components, and atomically engineered metallic alloys for radiation-tolerant nuclear energy systems. Hybrid nanomaterials have had the major (almost 35%) share of the nanotechnology industry focus, again certifying the importance of this technological field.

The progresses in nanomaterials are due to the cross-disciplinary efforts from scientists and engineers with a wide range of expertise in Physics, Chemistry, Biology, Mathematics, as well as Mechanical, Chemical, Materials, and Computer Engineering. The chapters in this book offer a strong flavor of the interdisciplinary nature of the research in nanomaterials. The six chapters of this book give detailed description of advanced nanomaterials including their synthesis, testing, and properties that are strongly associated with fundamental novelties in molecular mechanisms. The authors, who are nationally and internationally renowned experts in nanomaterials, and their corresponding chapters provide an overview to the uninitiated while a deeper understanding and knowledge to readers with specialized interests. The chapters are organized accordingly to the dimensionality of the principal nanomaterial discussed by the authors. In other words, Chaps. 1 and 2 discuss one-dimensional nanomaterial (carbon nanotube), Chaps. 3 and 4 discuss two-dimensional nanomaterial (graphene and its analogs), Chap. 5 presents advances in nanostructured three-dimensional oxide interfaces, and Chap. 6 relates to progress in nanofluidic materials.

In Chap. 1, M. Tehrani and P. Khanbolouki discuss the different aspects associated with the synthesis, processing, characterization, and properties of carbon nanotubes. As mentioned earlier, these materials have captured the interest of scientists over two decades and their commercial relevance has triggered their use in combination with other materials for designer structures. The authors begin from the very fundamentals of carbon chemistry and sequentially explain the engineering processes associated with developing these advanced materials. Their approach provides an easy read for the newcomers in this field, but also a comprehensive description of state-of-the-art technological aspects that would interest a specialist.

In Chap. 2, R. P. Sahu, I. K. Puri, and coauthors describe how carbon nanotubes in combination with magnetic nanoparticles have found significant interest and application in sensing opening new avenues for these nanomaterials. They meticulously discuss the synthesis procedures that include different methods of functionalization, and printable sensors that are magnetically and electrically responsive. The future of such advanced nanomaterials is extremely bright, to say the least, with unbound opportunities in supercapacitors, Lithium ion batteries as well as soft composite materials with significantly improved and controllable material properties.

In Chap. 3, S. Hu, S. Das, and H. Monshat present the progresses in the use of two-dimensional materials in electrochemical energy storage devices. The motivation for the research is rooted in the superior performance of these nanomaterials relative to the traditional materials employed in such applications. The discussion on supercapacitors builds organically from the previous chapter as the authors present the fundamental contributions of defects and functional groups to the material properties. An important aspect underlined in the outlook includes further research, regulation, and awareness on the effect of nanomaterials on environment and health.

In Chap. 4, O. Sanchez and coauthors summarize the knowledge on new elemental two-dimensional materials analogous to graphene. The elements in group four of the periodic table have found a niche interest in nanomaterials, the idea being mimicking the nanostructural aspects of carbon in silicon, germanium, and tin. Although the research is still in its infancy, the authors review the achievements in synthesis of silicene, germanene, and stanene, and their corresponding structural, electronic, and thermal properties.

In Chap. 5, TeYu Chien takes us to a different yet related territory of nanomaterial interfaces. The author discusses the novel phenomena that arise at the contact zones of dissimilar bulk complex oxide materials. Although the materials are of larger dimensions, the electronic properties are governed by the nanoscale features at the interfaces of these advanced materials. While advanced synthesis techniques can create these materials, the recent progress in cross-sectional scanning probe microscopy has facilitated the characterization of the typically challenging interfacial regions. The author provides a comprehensive background of the fundamentals of oxide materials, the properties at the interfaces that are investigated through state-of-the-art characterization, and next-generation applications that would benefit from these complex materials.

In Chap. 6, B. Ma and D. Banerjee show how dispersion of nanoparticles in fluids can enable improved properties for nanofluidic materials. The focus of the chapter is on synthesis procedures associated with "nanofluids." The considerable interest in these materials, albeit the controversies in the repeatability of their properties, requires a diligent synthesis protocol to facilitate desired functionality and material properties in such advanced fluids. The emphasis on the effect of synthesis conditions on the thermo-physical properties of nanofluids is particularly important for scalable manufacturing of these materials. In addition to providing fundamentals of nanofluids and their properties, the authors show how these futuristic liquids can be used for solar energy storage and power generation.

I sincerely express my appreciation to all the authors for sharing their knowledge and expertise through their respective chapters, and their impetus and interest towards the book. I thank Brian Halm, Michael Luby, and Nicole Lowary for extending this opportunity to me, and most importantly for their faith and patience through this process. I would also like to express my love and gratitude to my wife, Tanumita, for her love and support, and understanding my irregular work schedules, especially during the time that I was engaged with this book.

The support received through the National Science Foundation Award # CMMI-1404938 and the Summer Faculty Fellowships at the Air Force Research Lab in 2015 and 2016 are gratefully acknowledged.

Bethlehem, Pennsylvania, USA Ganesh Balasubramanian

Contents

About the Author

Ganesh Balasubramanian is an Assistant Professor of Mechanical Engineering and Mechanics at Lehigh University. Previously he was an Assistant Professor of Mechanical Engineering and (by courtesy) of Materials Science and Engineering at Iowa State University. He received his BME degree in Mechanical Engineering from Jadavpur University, India in 2007, his Ph.D. in Engineering Mechanics from Virginia Tech in 2011, and was a postdoctoral research associate in the Theoretical Physical Chemistry unit at TU Darmstadt in Germany till 2012. His research and teaching interests are in computational materials engineering, strongly disordered materials, and nanoscale transport and mechanics. Some of his recognitions include the ASEE Outstanding New ME Educator award, AFRL Summer Faculty Fellowship, Miller Faculty Fellowship at Iowa State, the Graduate Man of the Year and Liviu Librescu Scholarship at Virginia Tech, Young Engineering Fellowship from the Indian Institute of Science.

The original version of this book was revised. An erratum to this book can be found at DOI 10.1007/978-3-319-64717-3_7

Part I
One Dimensional Nanomaterials

Chapter 1
Carbon Nanotubes: Synthesis, Characterization, and Applications

Mehran Tehrani and Pouria Khanbolouki

1.1 Introduction

To understand the structure and properties of carbon nanotubes (CNTs), one needs to understand carbon chemical bonds. Carbon atoms can bond to other atoms via SP^3, SP^2, or SP covalent bonds. For example, carbon atoms are bonded via SP^3 bonds (sharing their four valence electrons with four other carbon atoms) in diamond. As a result, diamond is one of nature's hardest materials and has an ultrahigh thermal conductivity of ~1000 W/mK [1]. As shown in Fig. 1.1, graphite consists of stacked sheets of SP^2-bonded carbon atoms forming a honeycomb structure, that is, graphene. Graphene sheets interact with one another via weak van der Waals forces, and graphene layers slide easily over each other; thus, graphite is used as a lubricant. Graphene itself, however, is super strong and stiff—five times stiffer and more than 100 times stronger than steel [2]. A CNT can be visualized as a seamless cylinder of a rolled-up graphene sheet, as shown in Fig. 1.1b. CNTs are as strong and stiff as graphene and, like graphene, they possess very high electrical and thermal conductivity owing to the SP^2 carbon bonds. CNTs can be categorized by the number of concentric walls in the individual structures: single-, double-, and multiwalled CNTs. Single-walled CNTs (SWCNTs) can have diameters from 0.4 to 4 nm. Multiwalled CNTs (MWCNTs) are concentric shells of SWCNTs with intertube spacing of 0.34 nm, and their diameters typically range from 1.4 to 100 nm. CNTs cannot currently be continuously grown; however, individual nanotubes as long as half a meter have been demonstrated [3].

CNTs were discovered in 1991 by Iijima [5], and scientists have since been exploiting their extraordinary mechanical and physical properties. CNTs are a promising candidate for many applications, including—but not limited to—

M. Tehrani (✉) • P. Khanbolouki
Department of Mechanical Engineering, University of New Mexico,
Albuquerque, NM 87131, USA
e-mail: mtehrani@unm.edu; pouria@unm.edu

G. Balasubramanian (ed.), *Advances in Nanomaterials*,
DOI 10.1007/978-3-319-64717-3_1

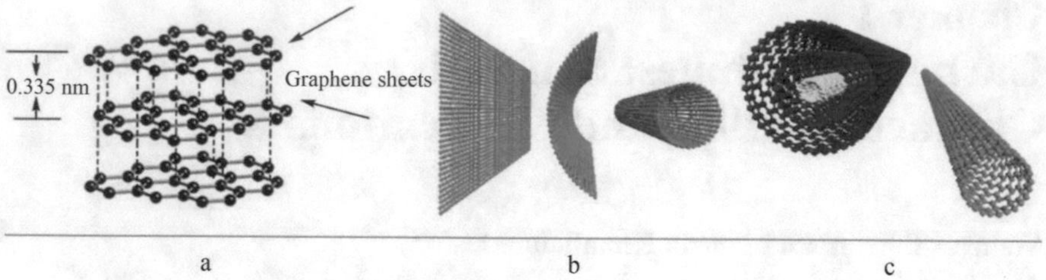

Fig. 1.1 Graphite, graphene [4] and different types of carbon nanotube: single walled, double walled and multiwalled carbon nanotubes. (**a**) graphite consists of stacked graphene sheets. (**b**) a single walled carbon nanotube can be regarded as a rolled graphene sheet. (**c**) A multi-walled and a single walled carbon nanotube

lightweight electrical wires, thermal interface materials, field-effect transistors, electrodes in energy storage and conversion devices, sensors and actuators, structural fibers and composites, and water desalination membranes [6].

1.2 Structure and Properties of CNTs

1.2.1 CNT Structure

As shown in Fig. 1.2, a CNT can be described by two directional indices (n, m). Directional indices (also described by the chiral vector, $na_1 + ma_2$) determine the rolling direction of graphene to form a nanotube, as well as the diameter, d, of a nanotube:

$$d = \frac{0.246\sqrt{n^2 + mn + m^2}}{\pi} \tag{1.1}$$

The angle between the chiral vector (n, m) and the horizontal vector (n, 0) in Fig. 1.2 is known as the chiral angle:

$$\theta = \sin^{-1} \frac{\sqrt{3m}}{2\sqrt{n^2 + mn + m^2}} \tag{1.2}$$

The chiral angle takes a value between 0 and 30° for different types of nanotubes. Specifically, an armchair nanotube ($n = m$) has a chiral angle of 30°, a zigzag nanotube ($n = 0$, or $m = 0$) has a chiral angle of 0°, and a chiral nanotube (any other n or m) has a chiral angle between 0 and 30°. The CNT bandgap varies by its chirality from 0 to 2 eV. The bandgap of semiconducting CNTs is inversely proportional to the nanotube diameter [7]. An SWCNT is a semiconductor if $n - m \neq 3i$ ($i = 1, 2,\ldots$) or metallic if $n - m = 3j$ ($j = 0, 1,\ldots$) [8]. For example, all armchair nanotubes are metallic, whereas zigzag nanotubes for which $n - m \neq 3j$ exhibit semiconducting characteristics.

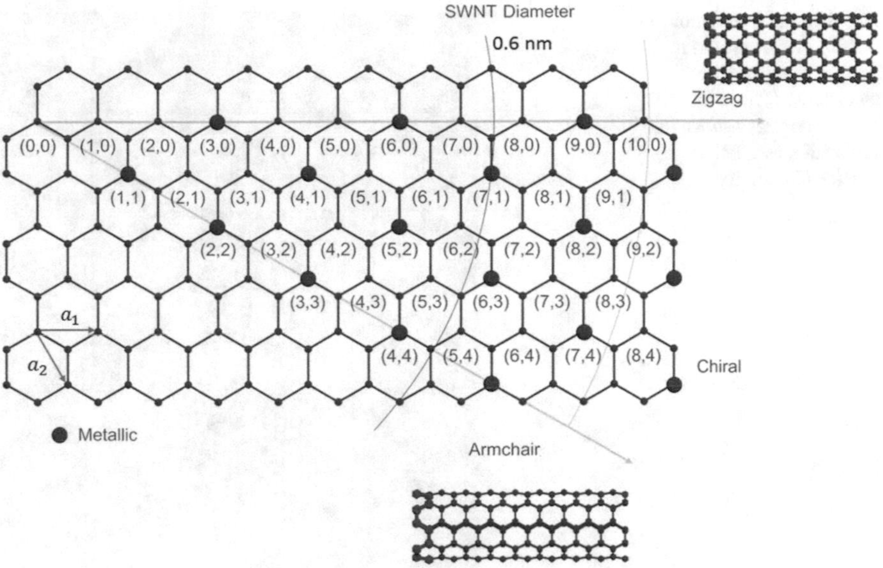

Fig. 1.2 Chirality table for single-walled carbon nanotubes (SWCNTs)

1.2.2 Electrical, Thermal, and Mechanical Properties

Ballistic transport is the ability of a material to transport electrons or phonons through its medium, with almost no resistance or scattering. Defect-free CNTs are very promising in electrical and thermal applications because of their ballistic transport ability over long lengths—with absence of electron/phonon scattering [9, 10]. As a result, they can carry high currents with almost no heating. The current-carrying capacity of metallic CNTs is expected to exceed 10^9 A/cm^2—four orders of magnitude higher than that of normal metals [7]. Semiconducting nanotubes display different sets of properties that make them highly desired for superior field-effect transistors [11, 12]. Another extraordinary property of CNTs is their thermal conductivity, ranging from 1000 to 6600 W/mK [13]. These values are 1–2 orders of magnitude higher than the thermal conductivities of metal, as well as natural diamond (2000 W/mK). Phonons are mainly responsible for heat conduction in CNTs [13]. SWCNTs have also exhibited superconductivity below 5 K [14].

While individual nanotube conductivities are surprisingly high, contact resistance between nanotubes, as well as between nanotubes and other materials, has made it difficult to translate individual nanotube properties to the micro- or macroscale. For example, for individual nanotubes with a conductivity of 3000 W/mK, one, two, and three-dimensional CNT networks have thermal conductivities of approximately 250, 50, and 3 W/mK, respectively [15]. The reduced thermal conductivities are due to the increased point contacts between individual CNTs in higher-dimensional networks, leading to increased interfacial thermal resistance

Fig. 1.3 Carbon nanotube (CNT) fibers support a lamp and supply the electricity. (Photo by Jeff Fitlow, courtesy of Matteo Pasquali's research group at Rice University)

(also called Kapitza resistance). Similarly, thermal conductivity of nanotubes can be degraded by defects in their structures [16].

Mechanical properties of individual CNTs and their interactions have been investigated both numerically and experimentally. They have a Young's modulus of up to 1.4 TPa, 20–30% elongation to failure, and a tensile strength higher than 100 GPa [17]. CNTs interact with each other via van der Waals interactions, and their interfacial shear strength is therefore only 0.24–1 MPa [18, 19]. In order to achieve a macroscale fiber made of CNTs, a highly packed, well-aligned, and optimally interconnected assembly of ultralong and highly pure CNTs should be achieved. Great progress toward this goal has been made; however, there is still much room for improvement. For example, fibers consisting of dense, highly aligned, and short (microns-long) nanotubes have been manufactured. Such fibers have achieved strengths lower than 1–2 GPa, stiffness of ~100 GPa, elongations smaller than 1–2%, electrical conductivity of 6.7×10^6 S/m, and thermal conductivity of 1230 W/mK [20]. Millimeters-long nanotubes have also been assembled into fibers. Nanotubes in such fibers are not well packed, nor are they highly aligned in the fiber direction; however, they have retained strengths of up to 8 GPa and electrical conductivity of 10^6 S/m [21]. Impurities, as well as voids, in these structures act as defects and degrade the mechanical performance of the resulting fibers. By engineering interfaces in CNT fibers, toughness values of up to 1000 J/g (10 times that of spider silk) have also been reported [22]. The combination of high mechanical and electrical/thermal properties makes CNT fibers beneficial in many applications. As an example, Fig. 1.3 shows two CNT fibers that are strong enough to hold a lamp while being electrically conductive enough to supply the electricity.

1.3 CNT Synthesis

Different synthesis methods have been developed for CNTs. Over the years, these methods have been improved and optimized to allow for control of the diameter and chirality, length, number of walls, crystallinity, and impurity. Other attempts have been focused on scaling up CNT production and continuous CNT growth. CNTs can be synthesized by several techniques such as arc discharge [23], laser ablation [24], sol-gel synthesis [25], the flame method [26, 27], and chemical vapor deposition (CVD), with CVD being the least expensive, easiest to scale, and most widely used one [28].

1.3.1 Carbon Arc Discharge

Iijima [5] utilized an arc discharge method to synthesize CNTs [23], leading to their discovery. A direct current (DC) is run between two vertical electrodes, which are placed in a reaction tube, and an arc discharge is generated between the electrodes. In this method, the thin anode electrode has small holes filled with a mixture of graphite and powder metals. The mixture gets vaporized by the discharge at high temperatures (1200 °C) and a flow of inert gas. SWCNTs grown using this method had an average diameter of 1 nm and were deposited on a collector downstream from the furnace. This method has the advantage of high yields but requires high energies and temperatures for sublimation of solid targets.

1.3.2 Laser Ablation of Carbon

Laser ablation is the process of evaporation or sublimation of a material heated by a laser beam. For synthesis of CNTs, Smalley et al. [24] used laser ablation of graphite sources where carbon atoms were assembled into the form of nanotubes.

1.3.3 Flame Method

In this method, a hydrocarbon reacts with an oxidizer to produce a precursor mixture that gets deposited on catalyst particles as CNTs. The growth substrate is positioned inside the flame, and the flame provides the required energy for the process [29]. It has been shown that the precursor in a flame synthesis process consists of oxygen and hydrogen, which have been speculated to have a positive influence on CNT growth in other methods [26, 27]. Fluid dynamics, mass transfer, and heat transfer each play a complex role in flame synthesis of CNTs [30].

1.3.4 Chemical Vapor Deposition

Its low cost (due to its relatively low operating temperatures), ability to control nanotube diameter and length, and scalability have made CVD the most readily available method for CNT synthesis. In CVD, nanotubes are grown from a catalyst particle. The catalyst is either on a substrate or formed in a gas flow inside a furnace. In particular, the precursor gases decompose, and CNTs are deposited on the catalyst particles. The precursor gases are usually a source of carbon (such as alcohols [31], aromatic compounds, or hydrocarbons) mixed with hydrogen and an inert gas. The most frequently used carbon sources in this process are ethylene, acetylene, methane, carbon monoxide, and ethanol. Through the years, many additives have been introduced to promote growth (such as water, hydrogen, and oxygen) or to dope the CNTs (such as nitrogen [32], phosphorous [33], and boron-bearing gases [34]). Many research groups have come up with variations of CVD apparatus to achieve more controllability of CNT growth. Some of these methods are point-arc microwave plasma CVD [34, 35], floating catalyst CVD [36], alcohol-assisted CVD [37, 38], molecular beam CVD [39], hot filament CVD (cold wall) [40, 41] oxygen-hydrogen-assisted CVD [42], water-assisted CVD (WACVD; a super growth method) [43, 44], fast-heating CVD [45], electric field CVD [46], low-pressure CVD [47], fluidized bed CVD [48], plasma-enhanced CVD [49], horizontal tube CVD, and a rotary tube furnace [50] (generally used for mass production). Among these, WACVD seems to be a cost-effective method for large-volume production of high-quality CNTs.

Although many parameters affect CNT growth, the catalyst is known to have a significant impact. Thus, preparation of the catalyst becomes of great importance for controlling the final CNT structure. For growing vertically aligned CNTs on a substrate, catalyst nanoparticles should be prepared prior to CNT growth. A variety of methods have been tried for preparing catalysts on substrates, such as the sol-gel technique [51], reduction of precursors [51], evaporation of a solution on the substrate, self-assembly (micellar solution or the reverse micelle method) [51], metal organic CVD [47], dip coating [52], atomic layer deposition [53], spin coating, electroplating of nanoparticles from a salt solution, contact printing of nanoparticle solutions, physical vapor sputtering, and, finally, evaporation techniques [54]. Reducing the two-step process of preparing the catalyst substrate and synthesis of CNTs to a one-step process can significantly improve the cost and rate of production. Reduction and breaking of precursor films into nanoparticles appears to be a promising approach for this purpose. Catalyst particles can also be generated in the growth tube. For example, in floating catalyst CVD, both the catalyst gas mixture and precursors are simultaneously introduced into the growth chamber. Catalyst nanoparticles are formed in the vapor phase, and nanotubes are grown from the catalysts and subsequently collected downstream.

Iron, cobalt, nickel, and their alloys have been extensively used to grow different types of nanotubes (i.e., single-, double-, and multiwalled CNTs). CNT synthesis with copper [55], gold [56], gadolinium, palladium [57], platinum [58], iridium [59], silver [60], rhenium, tungsten [61], yttrium [62], and molybdenum [63] cata-

Table 1.1 Comparison between primary methods of carbon nanotube (CNT) growth

Method	Yield	Quality	Purity	Temperature (°C)	Comments
Arc discharge	Low	Low	Low	~4000	
Laser ablation	Low	High	Medium	~1200	
Flame	Low	High	Medium	~1500	Highest growth rates achieved
CVD	High	Medium	Medium	100–1200	Greatest lengths achieved; more adaptable to a variety of structures

CVD chemical vapor deposition

lysts has also been demonstrated. Some elements have been used for alloying, such as molybdenum, magnesium, germanium, silicon, and even carbon [63, 64]. CNT growth using oxides of some of these elements has also been demonstrated [65, 66].

The advantages and disadvantages of different CNT synthesis approaches are summarized in Table 1.1.

Grown nanotubes contain different amounts of impurities in the form of catalyst particles or amorphous carbon. Postprocessing and purification steps are therefore usually performed to increase their purity. Such nanotubes are usually short and highly entangled. CNTs as long as centimeters can be grown using CVD while keeping their alignment, as shown in Fig. 1.4. To better illustrate nanotubes synthesis, CVD growth of vertically aligned CNTs is discussed here.

The growth substrate is usually a silicon wafer with hundreds of nanometers of thermally grown SiO_2 on top. If desired, nanotubes can be grown only in predetermined regions of this wafer. To this end, photolithography techniques can be used to create a pattern on the wafer. For example, a photoresist polymer is spin coated on the wafer and, by means of photolithography, a pattern is generated on it. The pattern acts as a mask, covering certain areas, while allowing access to others where the catalyst will be deposited and nanotubes will be grown. A buffer layer (e.g., 50 nm of Al_2O_3) and a catalyst film (e.g., 2 nm of Fe) is deposited on the wafer using a sputtering technique. The photoresist is then removed in the lift-off process. This wafer is ready for CNT growth.

CVD systems consist of a temperature/pressure-controlled furnace with a multi-gas delivery system. An example of a CVD system is shown in Fig. 1.5, where high-purity ethylene, argon, and hydrogen are introduced through mass flow controllers (MFCs). The downstream of the furnace may be connected to a double bubbler system with oil as a trap.

The furnace is first purged with a mixture of hydrogen and argon. The catalyst film is then reduced, and nanoparticles are formed in the presence of hydrogen and argon, and at high temperatures. The carbon source, e.g., ethylene, is then introduced into the system. By completion of the growth process, the ethylene feed is stopped, and the samples are removed after the furnace has cooled down. The following chart shows a representation of the process. A schematic of the growth process is shown in Fig. 1.6.

Fig. 1.4 (**a**) Vertically aligned carbon nanotubes (VACNTs). (**b**) Patterned VACNTs on a wafer substrate

1.3.5 CNT Growth Mechanisms

CNTs cannot be continuously grown yet. There have been many studies focused on underrating CNT growth and their self-termination growth mechanisms. In 2013, Zhang et al. [3] successfully grew a 550 mm–long CNT in 2 h using the CVD method. The precursors consisted of CH_4, H_2, and H_2O, and growth was carried out at 1100 °C. These results show the promising potential of CVD for achieving higher yields in manufacturing CNTs.

Recently, it was shown that with the traditional WACVD process, CNT forests can be grown to up to 2.17 cm [67]. Previously it was reported that water can prolong the lifetime of the catalyst nanoparticles [68]. Of course, a combination of parameters is involved in growing ultralong CNT forests, but, in theory, their synthesis can be continued for as long as the catalyst nanoparticles preserve their activity. In addition, the catalyst lifetime depends on the synthesis conditions such as the temperature, pressure, annealing conditions, water vapor concentration, and carbon precursor.

Without understanding the mechanism of the CNT synthesis, it is almost impossible to improve on the current methods to achieve more uniform structures with higher yields. Carbon dissolution–diffusion–precipitation [69] (similar to the vapor–liquid–solid process) is proposed to be one of the mechanisms involved in CNT growth, and some people believe there is more than just one mechanism involved in this process. The decomposition of the gaseous mixture is believed to happen mainly on the surface of catalytic nanoparticles, and formation of the CNTs occurs after diffusion of carbon atoms inside the catalysts. Thus, saturation of the nanoparticles is believed to be one of the mechanisms involved in termination of the process. Nucleation of the CNT base or cap is believed to happen adjacent to catalyst particles; then comes the growth process, which can be continued for several

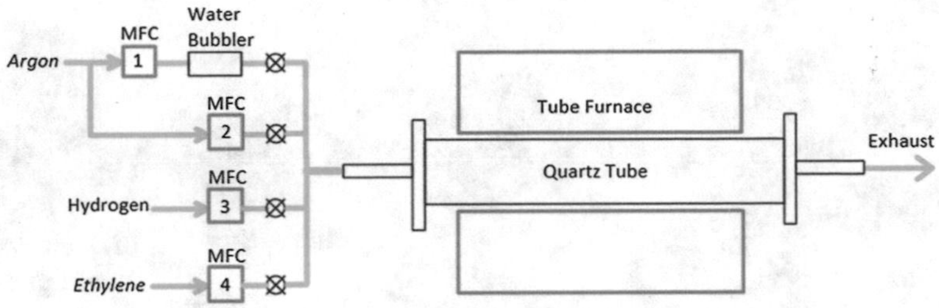

Fig. 1.5 Representative chemical vapor deposition (CVD) setup consisting of mass flow controllers (MFCs)

hours or days, depending on the process; then suddenly the growth stops. By understanding the termination process, we would be able to more efficiently control the growth of CNTs, and it has been connected to many factors in the process.

A plethora of studies have shown that deactivation of the catalyst occurs gradually [35, 70–72] or suddenly [73–76]. An abrupt decrease in the growth rate has also been shown in various studies [77–81]. Some of these studies discussed their hypotheses for growth termination. However, a universal explanation for this phenomenon would be more convenient if such a mechanism exists that can be applied to different procedures.

We can account for the following factors individually or in combination as suggested mechanisms of the VACNT growth rate decrease and growth termination:

- The diffusion rate of the carbon feedstock into the CNT forest [42, 82]
- The diffusion rates of the carbon in and on the catalyst nanoparticles [83]
- Catalyst poisoning and formation of carbon structures and oxides on or in catalyst nanoparticles [76, 84, 85]
- Ostwald ripening [86]
- Subsurface diffusion [87]
- CNT wall surface diffusion [88]
- Structure failure and van der Waals interactions of CNTs or covalent interaction of dangling bonds [89]

1.4 CNT Characterization

1.4.1 Individual CNTs

Nanotubes are usually characterized using scanning and transmission electron microscopy (SEM and TEM), Raman spectroscopy, and ultraviolet-visible near-infrared (UV-Vis-NIR) spectroscopy techniques. SEM is used to measure CNT length and morphology. TEM is used to determine the diameter and crystallinity of nanotubes. Raman spectroscopy is also an effective tool to study the structure of

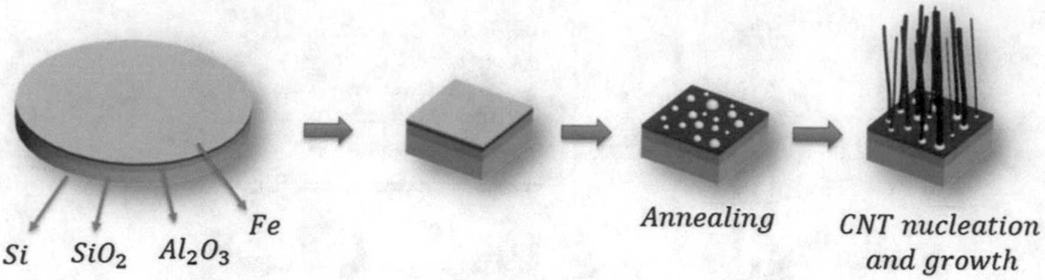

Fig. 1.6 Catalyst layer annealing, nucleation, and growth

CNTs by offering a quantitative measure of the CNT doping, crystallinity, and nanotube diameter. The ratio of the Raman D- to G-band (I_D/I_G), as shown in Fig. 1.7, is indicative of the CNT crystallinity, and the radial breathing mode can be correlated to the nanotube diameter and therefore its metallic or semiconducting nature [90]. If CNTs are dispersed in water, their UV-Vis-NIR absorption can be used to determine the ratio of metallic to semiconducting CNTs [91, 92]. A UV-Vis-NIR spectrum of well-dispersed semiconducting nanotubes is shown in Fig. 1.7, where the corresponding metallic peaks are missing.

1.4.2 CNT Structures

CNT assemblies possess a hierarchical structure across multiple length scales [93, 94]. In the simpler case of aligned CNT structures, as shown in Fig. 1.8, CNT structures (macroscale) in general consist of four building blocks at different length scales: (1) individual CNTs (~1–100 nm, nanoscale); (2) bundles (hundreds of nanometers, nanoscale)—several tightly bound side-by-side nanotubes; (3) fibrils >1 micron, mesoscale)—several loosely bound side-by-side bundles; and (4) different cross-linking blocks. The morphology and connectivity of these hierarchical structures determine the mechanical and physical properties of the CNT architectures. Resolving the structure across different length scales is crucial for understanding the behavior of CNT structures.

Monitoring of CNT structures in solutions and in dried forms can be realized using both real space imaging (e.g., light or electron microscopy) and reciprocal space analysis (e.g., light, X-ray, or neutron scattering) [95]. The former is a very powerful technique to elucidate structures and morphologies on the nano- and micron-length scales. Microscopy techniques are difficult to apply to solutions and provide only a two-dimensional slice of CNT morphology in the dry form that may be different from it bulk. Also, microscopy does not contain sufficient statistical information of the three-dimensional structure of the material, as it probes only very small regions. Scattering, on the other hand, gives ensemble structural information

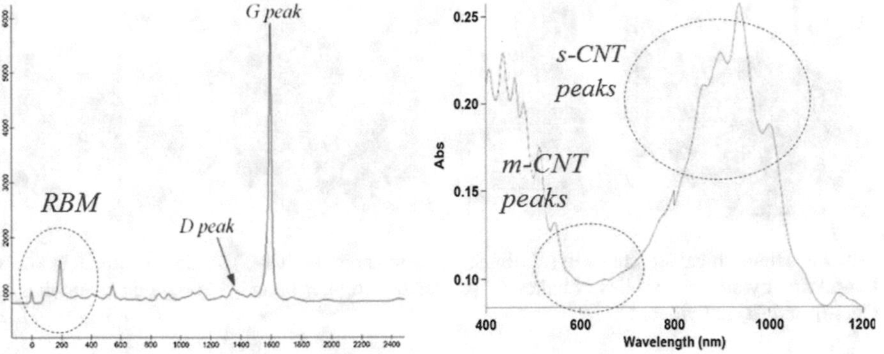

Fig. 1.7 *Left*: Intensity vs. Raman shift (cm^{-1}) spectrum for single-walled carbon nanotubes (SWCNTs) showing the radial breathing mode (RBM) and D and G peaks. *Right*: Ultraviolet-visible near-infrared (UV-Vis-NIR) spectrum for highly enriched semiconducting carbon nanotubes (s-CNTs). The metallic carbon nanotube (m-CNT) peaks are almost nonexistent, while the s-CNT peaks are sharp

averaged over relatively large volumes of the bulk sample, carrying detailed quantitative information [96–98]. Microscopy and scattering are, however, complimentary and can resolve the structure of CNT structures at the nano- and microscales (1 nm < probe size < hundreds of microns).

In particular, the size and alignment of CNTs, bundles, and pores inside a fibril will be extracted.

Nanotube alignment and packing most significantly affect the mechanical, electrical, and thermal performance of CNT structures. CNT alignment can be quantified using X-ray diffraction and Raman spectroscopy [20, 99, 100]. Polarized Raman spectroscopy also provides a useful probe to qualitatively measure CNT alignment by comparing the G-band intensity for parallely versus perpendicularly aligned sample directions [20, 101–103].

1.5 CNT Processing

Most CNT synthesis methods produce nanotubes with major amounts of impurities. These impurities are metal catalyst particles and amorphous carbon. Postacid treatment is usually used to increase their purity at the cost of damaged nanotubes and an increased price. VACNTs grown on a substrate can achieve high purity levels

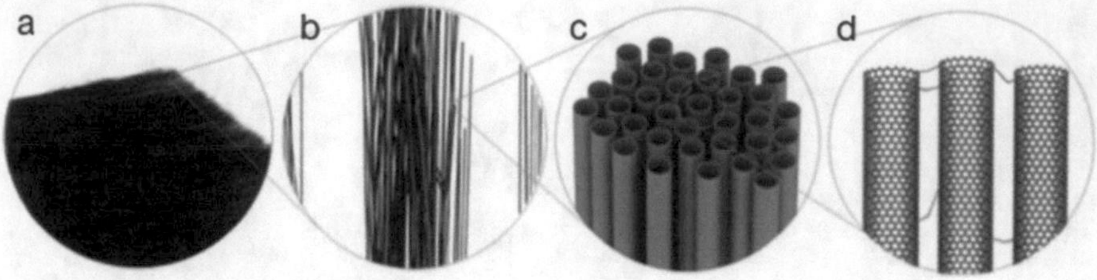

Fig. 1.8 Hierarchical structures in (**a**) aligned carbon nanotube (CNT) sheets; (**b**) fibrils (loosely bound side-by-side assemblies of bundles); (**c**) bundles (tightly bound side-by-side assemblies of CNTs); and (**d**) individual CNTs

without the need for postpurification. These nanotubes can be detached from the substrate, leaving the catalyst particles behind. Another approach to increase both the purity and crystallinity of CNTs is to subject them to elevated temperatures (up to 2500 °C) in an inert environment or a vacuum [104]. A crystalline defect has been shown to heal during such treatment, known as "graphitization."

CNTs can be grown from 100 nm to a few centimeters, depending on their intended application. The nanoscale size of nanotubes and their high aspect ratio have made their processing overwhelmingly difficult. As such, nanotubes form an entangled network of agglomerates and bundles that are not easy to disentangle. CNTs are therefore grown to less than tens of microns because of the difficulty in processing longer nanotubes. Such short nanotubes can be subsequently dispersed, using methods discussed here, and used in different applications. One of the most effective means to prevent agglomeration and control alignment in CNTs is to anchor one end of them (during synthesis) to a substrate, thus creating a stable structure [105, 106]. As mentioned earlier, this approach is used in CVD synthesis of VACNTs. CNTs can also be grown to a few millimeters in a floating catalyst CVD reactor and assembled into a fiber sheet form downstream. This approach is also effective in controlling agglomeration of nanotubes and results in aligned CNT fibers or sheets. Although the last two approaches are effective in preventing agglomeration, for many applications, short and entangled CNTs are processed into usable forms.

Chemical dispersion of CNTs, using surface modifiers (surfactants), can aid in CNT dispersion in different mediums. In general, CNTs have a smooth and nonreactive surface that does not interact with most solvents. Surfactants are usually amphiphilic molecules that wrap around or attach to nanotubes, thus enabling their dispersion. An amphiphilic has a hydrophilic polar head group and a hydrophobic tail group. The type of surfactant (cationic, anionic, nonionic, zwitterionic) is based on the head group charge [107]. For example, Triton X-100 is a nonionic surfactant used for dispersion of CNTs in aqueous solutions and can potentially enhance the bonding of CNTs to epoxy matrices [108, 109]. The surfactant forms a weak bond to the outer surface of the CNTs and allows for dispersion of the CNTs and separation of bundles into individual CNTs. Figure 1.9 depicts how a surfactant attaches to the outer walls of the CNTs.

The right amount of surfactant is needed to get the best dispersion. This amount is called the critical micelle concentration (CMC). Micelles are the self-organization of the surfactant molecules into small bundles. The CMC is the point at which the surfactant can adequately coat the tubes to disperse the bundles into individuals without forming micelles in the solution. An amount less than the CMC will not result in good dispersion and will leave bundles in the solution. Concentrations higher than the CMC will cause more bundles to form in the dispersion [108].

Another surface modification of CNTs is known as functionalization. There are different types of functionalization, including defect, covalent, and noncovalent functionalization [110]. The carboxyl functionalized nanotube is one of the most famous ones. Processes used to remove impurities in synthesized CNTs in return leave defects on the tubes in the form of -COOH groups (Fig. 1.10), [110]. Carboxyl groups alleviate van der Waals attractions between CNTs that cause bundling and can form covalent bonds to polymer matrices. The carboxyl groups that form on the CNTs are most commonly on the open ends, more so than on the outer walls, because of the higher concentration of defects on the ends. These defects make for better reactivity with the oxidation process [111].

Physical dispersion of CNTs is another effective approach for CNT dispersion. Physical approaches (Fig. 1.11) include a variety of methods: bath sonication, tip sonication, shear mixing, ball milling, and many others [110, 112, 113]. Physical and chemical approaches are usually used together to achieve better dispersion. Shear mixing draws the solution into a mixing head and pushes the solution through a narrow space between the rotor and the stator walls, shearing CNTs into uniform particles. The shear mixing does not damage the CNTs; it only breaks their agglomerates down into a uniform size. Once these agglomerates are broken down, the shear mixing process cannot aid in further dispersion and individualization of nanotubes. The product of shear mixing therefore contains bundles of CNTs, which can include hundreds to thousands of entangled CNTs. Usually, ultrasonication is needed to complete the dispersion of CNTs in a solution.

Tip sonication—compared with bath sonication—allows for a more focused and direct form of sonication. The tip sonicator has three major parts: the generator, converter, and probe/horn. Tip sonication is a direct form of sonication, where the probe is inserted directly into the solution. The probe vibrates while the tip expands and contracts during operation. The amount of expansion and contraction of the tip is the amplitude of the sonication [114]. This process creates the cavitation that is indicative of the sonication process. The cavitation bubbles create a high-energy stress wave upon bursting, which breaks down and unzips nanotubes from their bundles. When dispersed, nanotubes can rebundle if not stabilized. Bath sonication is more of an indirect form of sonication, where the sample is placed inside a water bath. The outer walls of the water bath cover a generator, which creates the sonication energy. Bath sonication is less powerful than tip sonication but is a technique that can process larger sample sizes and is less likely to damage nanotubes. Bath sonication does not produce the best CNT dispersion possible. The sonication energy is strong enough to break up some CNT bundles but not strong enough to fully individualize them (unless over extended periods of time and in very dilute CNT solu-

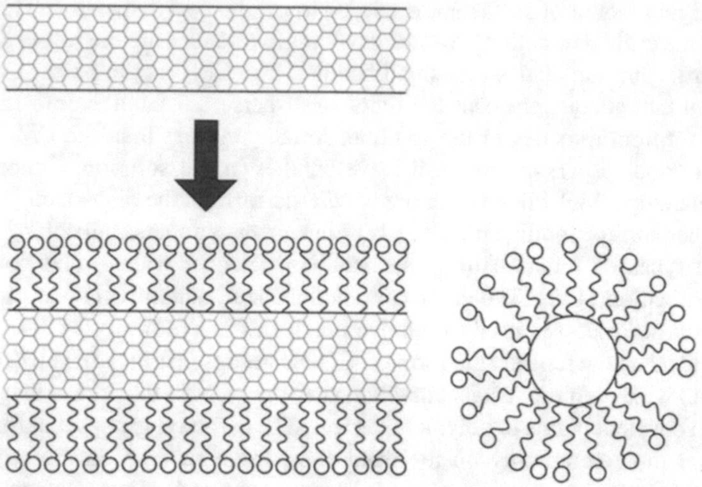

Fig. 1.9 Surfactant on the walls of a carbon nanotube (CNT): side view (*left*) and cross-sectional view (*right*)

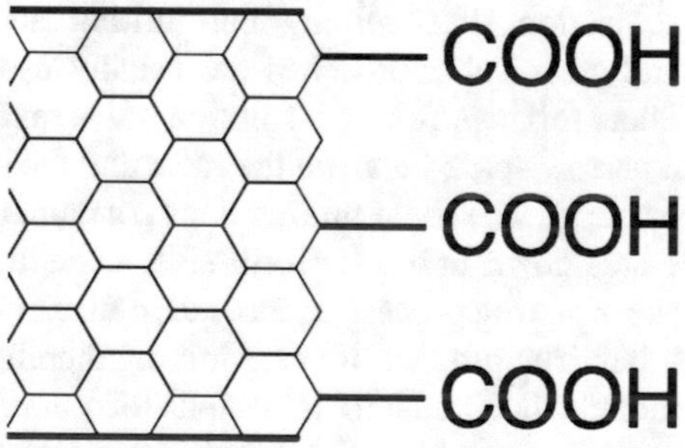

Fig. 1.10 -COOH groups on the tube ends of a carbon nanotube (CNT)

tions). However, it can prevent separated nanotubes from reagglomerating, while causing minimal damage to the CNTs. Both types of sonication cause damage to CNTs. Studies have shown that longer sonication times are beneficial to dispersion [115]. Intense sonication, however, causes too much damage and leaves CNTs unusable. A characterization step is required to check the dispersion of CNTs in the solution and ensure they are not damaged.

1.5.1 *Nanocomposites*

One of the first commercial applications of CNTs was electrically conductive polymer nanocomposites. The addition of small amounts of CNTs to polymers (usually less than 2–3 wt%) causes the electrical conductivity of the polymer to rise to 1–10 S/m (an increase of up to 15 orders of magnitude). When dispersed in

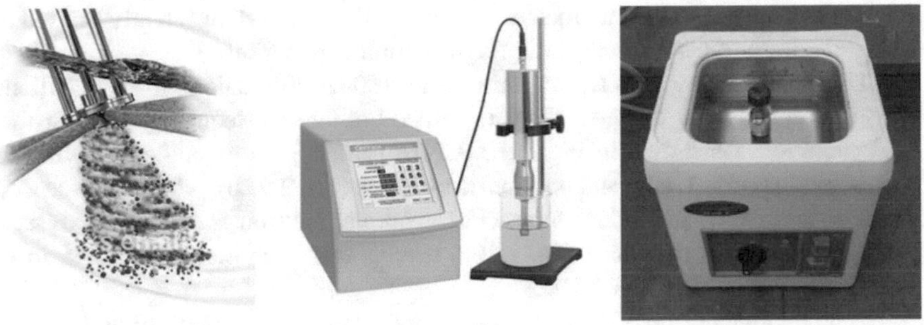

Fig. 1.11 *Left to right*: Shear mixer (http://www.silverson.com/us/products/batch-mixers/), tip (horn) sonicator (photo of Q500 Sonicator from Qsonica), and bath sonicator

polymers, nanotubes form an electrically conductive pathway. The better their dispersion is, the higher the resulting conductivity of the nanocomposites is. For example, a low percolation threshold (the minimum amount of nanotubes required to increase conductivity by at least five orders of magnitude) of 0.005 wt% for well-dispersed SWCNTs has been reported [116]. Similarly, CNTs can dramatically enhance the mechanical properties of polymers and even metals. This enhancement usually depends on the degree of dispersion, the CNT–matrix bonding that provides effective stress transfer, the intrabundle sliding within nanotubes ropes, and the processes used for CNT dispersion.

Because of the many different ways in which researchers are dispersing CNTs into polymer matrices, there have been varying—and sometimes conflicting—studies of the mechanical properties of the resulting nanocomposites. Various solvents, surfactants, physical dispersion methods, and processes for CNT dispersion have been investigated. Moreover, depending on their synthesis method, CNTs can have different properties. Therefore, the best dispersion processes are the ones that achieve the best properties of the nanocomposites. In general, adding more than a 2–3% loading of nanotubes to polymers makes their processing difficult, because of the significantly increased viscosity of the CNT–polymer mixture, and results in agglomeration of CNTs. Certain polymers can be used as both the matrix and the surfactant in polymer dispersion. When dispersed in a medium, CNTs can be filtered to form a CNT paper or "buckypaper." The properties of these papers are, however, orders of magnitude lower than those of individual nanotubes. For example, careful purification and assembly of the nanotubes into buckypapers, and even their impregnation with polymers to improve stress transfer, have resulted in Young's modulus values (<10 GPa) better than those of polymers (1–3 GPa) and far from those of metals or individual nanotubes [117, 118]. The improvement in the mechanical properties of polymers by addition of CNTs, although considered significant over the base properties of the neat polymer, is far from the theoretical predictions. Despite the great promise of CNTs for many engineering applications, there is still no scalable method to effectively disperse nanotubes and transfer stresses to them via a polymer matrix. These two limitations have greatly limited the widespread use of CNTs, especially for structural applications.

CNTs possess very high aspect ratios (e.g., 100–1000), making it very challenging to homogenously disperse them in a medium and subsequently prevent their agglomeration. As shown in Fig. 1.12, CNTs form highly entangled networks in both solutions (Fig. 1.12a) and polymeric matrices (Fig. 1.12b), which include CNT bundles rather than individual tubes. For example, neutron scattering analysis of nanocomposites, as shown in Fig. 1.12a, reveals that each bundle (as seen in the figure) consists of 50–100 CNTs.

Disentanglement of CNTs is possible only for very dilute solutions and requires both complicated chemical and physical dispersion approaches that damage CNT structures and break them into shorter tubes [90, 119–121]. As depicted in Fig. 1.13, proper chemical and physical dispersion techniques can result in relatively well-dispersed CNTs in a polymer matrix. There is, however, minimal control over how they assemble during drying or incorporation into a polymer matrix. The promise of CNTs would be delivered if they were individually dispersed in a matrix and pointing in desired directions [120]. The current techniques to disperse CNTs are, however, effective only at very low CNT loadings and, most importantly, do not provide a means to control their assembly [122–124].

1.5.2 CNT Fibers

Much effort has been devoted to producing aligned CNT structures (fiber, yarns, or sheets) in order to translate individual nanotube properties to the macroscale [21]. When normalized by weight, bulk CNT fibers surpass the properties of the best materials known to humans. For example, CNT fibers have exhibited specific (divided by weight) strength and stiffness values that are 10- and 4-fold higher than those of the strongest (T1000G) and stiffest (M60J) commercial carbon fibers [125], respectively. They have also achieved specific electrical and thermal conductivities in excess of the properties of copper [126, 127]. Three distinct and scalable routes to the manufacturing of such macroscale structures have been developed. The first approach—wet processing [126]—involves extrusion of a premade CNT solution through a spinneret into a coagulation bath, forming dense CNT fibers. CNTs are dispersed either in a strong acid (sulfuric and chlorosulfonic acids are the only nanotube solvents) or in other aqueous polymer solutions. Only fibers produced from acid solutions have better alignment and packing. The other two are solid-state methods—involving dry processing [128]—to spin CNT yarns or sheets directly from a floating catalyst CVD reactor or from a pregrown VACNT array. Both dry and wet processing techniques have obtained exemplary properties.

Wet-processed fibers (spun from a solution) consist of dense, highly aligned, and mostly catalyst-free nanotubes. Such fibers have achieved strengths lower than 1–2 GPa, stiffness of ~100 GPa, and elongations smaller than 1–2%. These fibers nevertheless have the highest reported electrical and thermal conductivities among

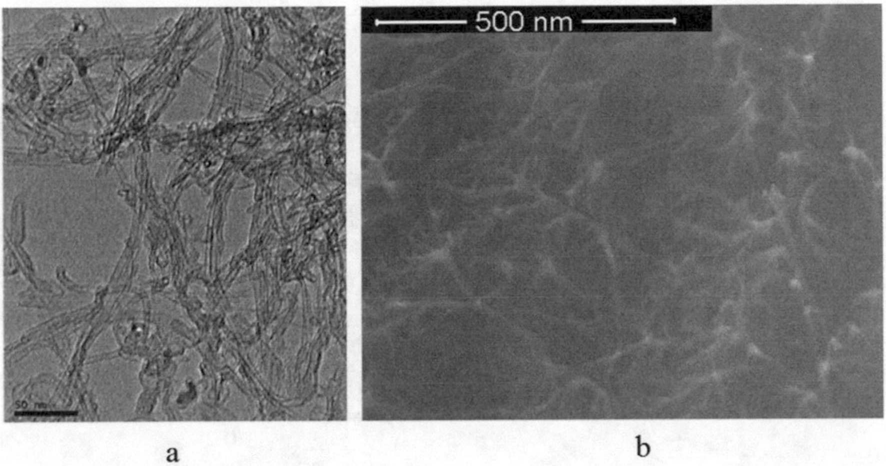

Fig. 1.12 (**a**) Carbon nanotube (CNT) bundles exhibiting considerable entanglement and agglomeration. (**b**) Scanning electron microscope (SEM) image of a nanocomposite containing 10 wt% CNTs. Bundles of CNTs form a random 3-dimensional (3D) network of CNTs in a polymer. Each bundle in this image consists of approximately 30–100 CNTs

CNT fibers. Only microns-long nanotubes can be processed using the existing wet processing methods [20]. Dry processing from floating catalyst CVD achieves fibers and sheets with suboptimal packing (voids) and alignment, and leaves catalyst and amorphous carbon impurities in the final structure [21, 129]. Despite these shortcomings, this method remains one of the most promising approaches to produce ultrastrong (up to 8 GPa) and ultraconductive (10^6 S/m) CNT structures. This is due to the long, high-quality, and small-diameter nanotubes synthesized using this method. The last method—dry processing from VACNTs—requires a specific type of VACNT: the so-called spinnable VACNT. Twisting, tension, or liquid shrinking during drawing have been shown to improve the alignment of CNTs in these fibers. They have achieved strength of up to 2 GPa and elongation of up to 7%. Although these properties make CNT fibers superior to structural fibers or materials, they are still an order of magnitude lower than individual CNT properties. There are many factors that contribute to this: (1) poor CNT packing and alignment; (2) CNT end junctions, which act as defects in aligned CNT structures; and (3) defects (catalyst and voids) that limit their electrical/thermal and mechanical performance [126, 130]. Continuous growth of CNTs or manufacturing techniques that allow for packing and alignment of impurity-free and ultralong nanotubes are needed to fully unlock CNT potential. Engineering of CNT–CNT and CNT–polymer interfaces for improved stress transfer and electron/phonon transport are equally important to achieve superior CNT structures.

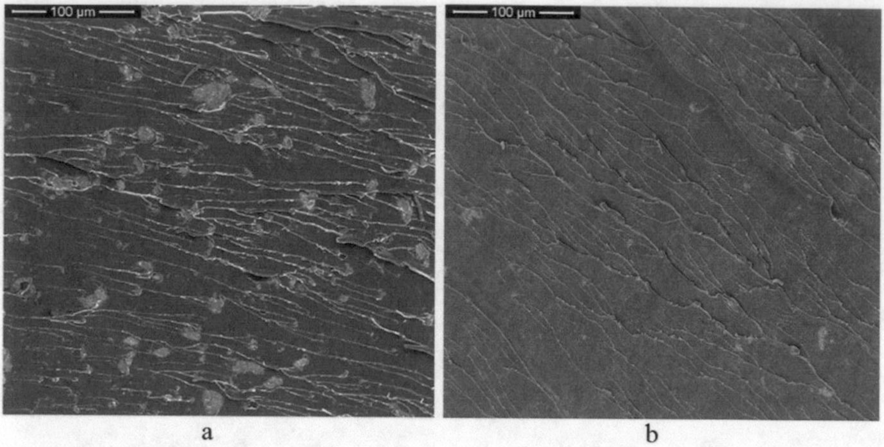

Fig. 1.13 Fractography micrographs of (**a**) a nanocomposite exhibiting carbon nanotube (CNT) agglomerates and (**b**) good CNT dispersion

1.6 Semiconducting CNTs

Synthesis methods generally produce inhomogeneous mixtures that contain CNTs with various chiralities and diameters. The ratio is usually only one-third metallic CNTs (m-CNTs) to two-thirds semiconducting CNTs (s-CNTs). While m-CNTs are required for their high electrical conductivity, s-CNT are highly desired for superior field-effect transistors [11]. Methods for achieving s-CNTs have been widely developed and studied [131–138]. This is due to the high demand for smaller transistors and the high potential of CNTs to outperform current transistors [11]. Three approaches for producing s-CNTs exist: (1) direct synthesis [132, 133, 139, 140]; (2) postsynthesis separation of nanotubes [141, 142]; and (3) postsynthesis elimination of metallic nanotubes [133, 143–145]. Postsynthesis separation of CNTs can enrich their semiconducting or metallic portions by over 99%; however, current methods are intensely labor intensive, are expensive, are only applicable to microns-long nanotubes, and produce very small quantities of CNTs [142]. The most common methods for CNT separation involve gel-chromatography separation, density-based separation, hydrophobicity-based separation, and aqueous two-phase separation (ATP) [141, 142]. At similar diameters, metallic CNTs are more reactive than semiconducting ones because of their smaller ionization potential [133]. For example, gas-phase plasma hydrocarbonation selectively etches metallic CNTs, leaving semiconducting CNTs relatively unchanged [146]. Certain gases, catalysts, and substrates have also been shown to favor the growth of semiconducting CNTs [131–138]. Catalysts with active oxygen groups, for example, etch metallic CNTs and result in CVD-grown CNTs with high semiconducting purity.

Table 1.2 Potential applications and challenges of carbon nanotubes (CNTs)

Field of application	Application	Type of CNT/processing method	Advantages	Challenges
Composites	Conductive fillers in polymers for: Automotive fuel lines and filters that dissipate electrostatic charges Electrostatic-assisted painting of mirror housings EMI-shielding packages Wafer carriers for microelectronics [6]	MWCNT or SWCNT–polymer composites	Small CNT amounts can create an electrically conductive network	CNT–matrix interactions and CNT dispersion are not optimal
	Load-bearing composites for baseball bats, tennis racquets, bike frames, and wind turbine blades [151, 152]	MWCNT powders mixed with epoxy resins	Improved stiffness, strength, toughness, and damping [153]	Uniform dispersion in matrix CNT–matrix adhesion CNT intratube and intrabundle sliding [153]
	Fuzzy fibers for lightning strike protection, deicing, and structural health monitoring for aircraft [151]	Aligned CNT forests	Mechanical, impact, and damping improvements [106, 124]	Direct growth of CNT on fibers can degrade fiber properties
	Superconducting wires, battery and catalytic nanofibers for fuel cell electrodes, and self-cleaning textiles [154]	Coating forest-drawn CNT sheets with functional powder	Weavable, braidable, knottable, and sewable yarns	
	Metal composites for automotive and aerospace industries [155]	Aluminum–MWCNT composites	Mechanical properties Weight reduction	Price Dispersion

(continued)

Table 1.2 (continued)

Field of application	Application	Type of CNT/processing method	Advantages	Challenges
	Replacement for halogenated retardant [156]	Mixing with resins	Flame retardant	
	Thermal interfaces	Combined CNT–graphene networks or vertically grown CNT arrays	Mechanically robust and very high thermal conductivity	High thermal resistance at interfaces
	Anticorrosion coatings for metal electrical pathway for cathodic protection in the marine industry [157]	MWCNT-containing paints and silicon-based coatings	Processability in paints	
	Transparent flexible conductive films for displays, touch screens, heated windows, and sidewalks [6, 158]	CNT solutions	Cost-effective nonlithographic methods and flexibility	Still possess higher resistance than ITO with equal transparency
	CNT thin-film heaters [6]			
Microelectronics	Field-effect transistors [159]	Semiconducting SWCNTs	Low electron scattering Favorable bandgap	Insufficient control over CNT diameter, chirality, and placement for large areas
	Tunnel field-effect transistors [160]	Ultrathin CNT semiconducting films	Energy reduction Quantum-mechanical band-to-band tunneling	
	CNT thin-film transistors in OLED displays [161]	CNT 2D thin films	High mobility Low-temperature, nonvacuum deposition methods	Better understanding of CNT surface chemistry is required

Field of application	Application	Type of CNT/processing method	Advantages	Challenges
5:	Microelectronic interconnects	Tightly packed metallic CNTs	Low electron scattering High current-carrying capacity High resistance to electromigration	Tightly packed, low-contact-resistance, low-defect metallic CNTs are required
	Field emission electron source for flat panel displays, X-ray and microwave generators, and lamps and gas discharge tubes	Screen-printed nanotube paste	Low power consumption High response rate	Competitive price alongside technical challenges
	Electromechanical switches and actuators [162]	Tangled CNT thin films	High temperature resistance Low operation voltage Mechanical properties	Achieving CNT alignment in sheets and fibers
	Nanoscopic tweezers [163]	CNTs attached to independent electrodes		
Energy storage and environment	Batteries [164, 165]	MWCNT powders	Electrical connectivity and mechanical integrity Weight reduction High reversible component of storage capacity at high discharge rates	Price Hysteresis
	Supercapacitors for hybrid and electric vehicles [166]	Forest-grown SWCNTs	High accessible surface area in porosities of arrays	Price
	Fuel cells [167, 168] and organic solar cells [169]	Individual SWCNTs as electrodes	1D	Controversial results from different research groups
	Electrodes in photovoltaics [170]	SWCNTs	Small Fermi velocity and low dielectric constant	

(continued)

Table 1.2 (continued)

Field of application	Application	Type of CNT/processing method	Advantages	Challenges
	Portable filters for water purification [171–173]	Tangled 2D CNT networks Aligned CNTs for water flow through the interior of CNTs VA-DWCNTs	Mechanical and electrochemical properties Higher gas and water permeability [171]	Need for very-small-diameter SWCNTs or DWCNTs for salt rejection at seawater concentrations [174]
Biotechnology	Biosensors Diagnostic and therapeutic applications Fluorescent molecular imaging [175] Localized heating using near-infrared radiation Ink-jet-printed test strips for estrogen and progesterone detection [6] Microarrays for DNA and protein detection [6] NO_2 and cardiac troponin sensors [176]	Conjugation of targeting ligands to SWCNT tags [177] DNA-SWCNTs [175]	Large changes in electrical impedance [178] and optical properties (photostable near-infrared emission for prolonged detection through biological media and single-molecule sensitivity) [179] Dimensional and chemical compatibility with biomolecules High sensitivity to the surrounding environment High surface area, semiconducting behavior, bandgap fluorescence, and strong Raman scattering spectra [177]	High selectivity and differentiation between adsorbed species in mixtures is needed Toxicity is still a concern
	Photo acoustic molecular imaging [180]	SWCNTs conjugated to cyclic Arg-Gly-Asp (RGD) peptides		

Field of application	Application	Type of CNT/processing method	Advantages	Challenges
	Label-free electrical detection of nucleic acids [178]	SWCNTs	Size and structural advantages Cost effectiveness	Amplification is required
	DNA sequence detection [176]	CNT network field-effect transistors	Low complexity High accuracy	Toxicity concerns
	Gas [181] and toxin [182] detection sensors	Mixture of SWCNTs with copper [181] SWCNTs coated with chemoselective materials [182]	Highly sensitive resistance [181] and capacitance [182] changes for sub-ppm detections Price Preparation simplicity	
	In vivo drug delivery [177, 183]			More toxicity studies are required Prevention of undesirable accumulation, which may result from changing surface chemistry [6, 184]

1D one-dimensional, *2D* two-dimensional, *DWCNT* double-walled carbon nanotube, *EMI* electromagnetic interference, *ITO* indium tin oxide, *MWCNT* multiwalled carbon nanotube, *OLED* organic light-emitting diode, *ppm* part-per-million, *RGD* Arg-Gly-Asp, *SWCNT* single-walled carbon nanotube, *VA* vertically aligned

1.7 CNT Applications

Owing to their high surface area and physical properties, CNTs have found applications in many areas including—but not limited to—hydrogen storage and electrochemical devices, filed emission, thermoelectrics [147], tissue engineering and drug delivery [148, 149], and nanoelectronics and sensing/actuating [6, 7]. CNTs have also been used in polymer composites for lightning strike protection, deicing, structural health monitoring, and electromagnetic shielding [6]. Besides polymer composites, the addition of small amounts of CNTs to metals has been shown to improve both the tensile strength and the modulus [150]. Low densities, solution processability, chemical stability, high flexibility in bulk forms, and sensitivity to surrounding mediums for sensing applications are just a few of these characteristics that can lead to advances in electronics, composite materials, biotechnology, and energy storage. On the other hand, various challenges remain to be resolved for achieving this goal in CNT-based device fabrication. Selective growth of CNTs of a specified size, chirality, and placement is still a subject of ongoing research. Addressing the challenges of growth, sorting, and assembly of high-purity CNT structures is of great importance. Higher growth yields and control over packing densities are crucial for some applications. Provided that all of these goals are achieved, research also needs to be performed on the sustainability, environmental aspects, and life cycle of these newly developed devices. Moreover, alongside the experimental research and manufacturing developments, advances are needed in characterization techniques, with better understanding of the underlying theoretical relations of these new structures to device performance. The potential applications and challenges of CNTs are briefly summarized in Tables 1.1 and 1.2.

References

1. Olson JR, Pohl RO, Vandersande JW, Zoltan A, Anthony TR, Banholzer WF (1993) Thermal-conductivity of diamond between 170 and 1200-K and the isotope effect. Phys Rev B 47:14850–14856
2. Lee C, Wei XD, Kysar JW, Hone J (2008) Measurement of the elastic properties and intrinsic strength of monolayer graphene. Science 321:385–388
3. Zhang RF, Zhang YY, Zhang Q, Xie HH, Qian WZ, Wei F (2013) Growth of half-meter long carbon nanotubes based on Schulz-Flory distribution. ACS Nano 7:6156–6161
4. Naz A, Kausar A, Siddiq M, Choudhary MA (2016) Comparative review on structure, properties, fabrication techniques, and relevance of polymer nanocomposites reinforced with carbon nanotube and graphite fillers. Polym-Plast Technol Eng 55:171–198
5. Iijima S (1991) Helical microtubules of graphitic carbon. Nature 354:56–58
6. De Volder MFL, Tawfick SH, Baughman RH, Hart AJ (2013) Carbon nanotubes: present and future commercial applications. Science 339:535–539
7. Baughman RH, Zakhidov AA, De Heer WA (2002) Carbon nanotubes—the route toward applications. Science 297:787–792
8. Dekker C (1999) Carbon nanotubes as molecular quantum wires. Phys Today 52:22–28

9. Grado-Caffaro MA, Grado-Caffaro M (2008) On ballistic transport in carbon nanotubes. Optik 119:601–602
10. Li HJ, Lu WG, Li JJ, Bai XD, Gu CZ (2005) Multichannel ballistic transport in multiwall carbon nanotubes. Phys Rev Lett 95:086601
11. Javey A, Guo J, Wang Q, Lundstrom M, Dai HJ (2003) Ballistic carbon nanotube field-effect transistors. Nature 424:654–657
12. Min CY, Shen XQ, Shi Z, Chen L, Xu ZW (2010) The electrical properties and conducting mechanisms of carbon nanotube/polymer nanocomposites: A review. Polym-Plast Technol Eng 49:1172–1181
13. Han ZD, Fina A (2011) Thermal conductivity of carbon nanotubes and their polymer nanocomposites: A review. Prog Polym Sci 36:914–944
14. Tang ZK, Zhang LY, Wang N, Zhang XX, Wen GH, Li GD, Wang JN, Chan CT, Sheng P (2001) Superconductivity in 4 angstrom single-walled carbon nanotubes. Science 292:2462–2465
15. Prasher RS, Hu XJ, Chalopin Y, Mingo N, Lofgreen K, Volz S, Cleri F, Keblinski P (2009) Turning carbon nanotubes from exceptional heat conductors into insulators. Phys Rev Lett 102:105901
16. Berber S, Kwon YK, Tomanek D (2000) Unusually high thermal conductivity of carbon nanotubes. Phys Rev Lett 84:4613–4616
17. Shokrieh MM, Rafiee R (2010) A review of the mechanical properties of isolated carbon nanotubes and carbon nanotube composites. Mech Compos Mater 46:155–172
18. Filleter T, Yockel S, Naraghi M, Paci JT, Compton OC, Mayes ML, Nguyen ST, Schatz GC, Espinosa HD (2012) Experimental–computational study of shear interactions within double-walled carbon nanotube bundles. Nano Lett 12:732–742
19. Paci JT, Furmanchuk A, Espinosa HD, Schatz GC (2014) Shear and friction between carbon nanotubes in bundles and yarns. Nano Lett 14:6138–6147
20. Ericson LM, Fan H, Peng HQ, Davis VA, Zhou W, Sulpizio J, Wang YH, Booker R, Vavro J, Guthy C, Parra-Vasquez ANG, Kim MJ, Ramesh S, Saini RK, Kittrell C, Lavin G, Schmidt H, Adams WW, Billups WE, Pasquali M, Hwang WF, Hauge RH, Fischer JE, Smalley RE (2004) Macroscopic, neat, single-walled carbon nanotube fibers. Science 305:1447–1450
21. Lu WB, Zu M, Byun JH, Kim BS, Chou TW (2012) State of the art of carbon nanotube fibers: opportunities and challenges. Adv Mater 24:1805–1833
22. Shin MK, Lee B, Kim SH, Lee JA, Spinks GM, Gambhir S, Wallace GG, Kozlov ME, Baughman RH, Kim SJ (2012) Synergistic toughening of composite fibres by self-alignment of reduced graphene oxide and carbon nanotubes. Nat Commun 3:650
23. Parker DH, Wurz P, Chatterjee K, Lykke KR, Hunt JE, Pellin MJ, Hemminger JC, Gruen DM, Stock LM (1991) High-yield synthesis, separation, and mass-spectrometric characterization of fullerenes C60 to C266. J Am Chem Soc 113:7499–7503
24. Scott CD, Arepalli S, Nikolaev P, Smalley RE (2001) Growth mechanisms for single-wall carbon nanotubes in a laser-ablation process. Appl Phys A 72:573–580
25. Valizadeh M, Kazemzadeh A, Raisian M, Mohammadizadeh A (2007) Development of sol-gel process for synthesis of single-walled carbon nanotubes. Asian J Chem 19:1246–1250
26. Hu WC, Hou SS, Lin TH (2014) Analysis on controlling factors for the synthesis of carbon nanotubes and nano-onions in counterflow diffusion flames. J Nanosci Nanotechnol 14:5363–5369
27. Liu YC, Sun BM, Ding ZY (2011) Effect of hydrogen on V-type pyrolysis flame synthesis of carbon nanotubes. Adv Polym Sci Eng 221:545–549
28. Prasek J, Drbohlavova J, Chomoucka J, Hubalek J, Jasek O, Adam V, Kizek R (2011) Methods for carbon nanotubes synthesis-review. J Mater Chem 21:15872–15884
29. Diener MD, Nichelson N, Alford JM (2000) Synthesis of single-walled carbon nanotubes in flames. J Phys Chem B 104:9615–9620
30. Unrau CJ, Katta VR, Axelbaum RL (2010) Characterization of diffusion flames for synthesis of single-walled carbon nanotubes. Combust Flame 157:1643–1648

31. Yamamoto S, Tani K, Onaka Y, Takata Y, Suzuki S, Shibuta Y, Maruyama S, Kohno M (2007) Synthesis of single walled carbon nanotubes by laser vaporized catalytic chemical vapor deposition technique. ASME/JSME 2007 Thermal Engineering Heat Transfer Summer Conference 2:387–393
32. Xue RL, Sun ZP, Su LH, Zhang XG (2010) Large-scale synthesis of nitrogen-doped carbon nanotubes by chemical vapor deposition using a co-based catalyst from layered double hydroxides. Catal Lett 135:312–320
33. Campos-Delgado J, Maciel IO, Cullen DA, Smith DJ, Jorio A, Pimenta MA, Terrones H, Terrones M (2010) Chemical vapor deposition synthesis of N-, P-, and Si-doped single-walled carbon nanotubes. ACS Nano 4:1696–1702
34. Watanabe T, Tsuda S, Yamaguchi T, Takano Y (2010) Microwave plasma chemical vapor deposition synthesis of boron-doped carbon nanotube. Physica C-Superconductivity and Its Applications 470:S608–S609
35. Zhong G, Iwasaki T, Robertson J, Kawarada H (2007) Growth kinetics of 0.5 cm vertically aligned single-walled carbon nanotubes. J Phys Chem B 111(8):1907–1910
36. Xiang R, Luo G, Yang Z, Zhang Q, Qian W, Wei F (2007) Temperature effect on the substrate selectivity of carbon nanotube growth in floating chemical vapor deposition. Nanotechnology 18(41):415703
37. Li Y, Xu G, Zhang H, Li T, Yao Y, Li Q, Dai Z (2015) Alcohol-assisted rapid growth of vertically aligned carbon nanotube arrays. Carbon 91:45–55
38. Unalan HE, Chhowalla M (2005) Investigation of single-walled carbon nanotube growth parameters using alcohol catalytic chemical vapour deposition. Nanotechnology 16:2153–2163
39. Eres G, Kinkhabwala AA, Cui H, Geohegan DB, Puretzky AA, Lowndes DH (2005) Molecular beam–controlled nucleation and growth of vertically aligned single-wall carbon nanotube arrays. J Phys Chem B 109(35):16684
40. Xu YQ, Flor E, Kim MJ, Hamadani B, Schmidt H, Smalley RE, Hauge RH (2006) Vertical array growth of small diameter single-walled carbon nanotubes. J Am Chem Soc 128(20):6560–6561
41. Okazaki T, Shinohara H (2003) Synthesis and characterization of single-wall carbon nanotubes by hot-filament assisted chemical vapor deposition. Chem Phys Lett 376:606–611
42. Zhang G, Mann D, Zhang L, Javey A, Li Y, Yenilmez E et al (2005) Ultra-high-yield growth of vertical single-walled carbon nanotubes: hidden roles of hydrogen and oxygen. Proc Natl Acad Sci U S A 102(45):16141–16145
43. Yasuda S, Futaba DN, Yamada T, Satou J, Shibuya A, Takai H et al (2009) Improved and large area single-walled carbon nanotube forest growth by controlling the gas flow direction. ACS Nano 3(12):4164–4170
44. Wang X, Li Q, Xie J, Jin Z, Wang J, Li Y et al (2009) Fabrication of ultralong and electrically uniform single-walled carbon nanotubes on clean substrates. Nano Lett 9(9):3137–3141
45. Huang S, Woodson M, Smalley R, Liu J (2004) Growth mechanism of oriented long single walled carbon nanotubes using "fast-heating" chemical vapor deposition process. Nano Lett 4(6):1025–1028
46. Peng BH, Jiang S, Zhang YY, Zhang J (2011) Enrichment of metallic carbon nanotubes by electric field-assisted chemical vapor deposition. Carbon 49:2555–2560
47. Ikuno T, Katayama M, Yamauchi N, Wongwiriyapan W, Honda S, Oura K, Hobara R, Hasegawa S (2004) Selective growth of straight carbon nanotubes by low-pressure thermal chemical vapor deposition. Jpn J Appl Phys 43:860–863
48. Venegoni D, Serp P, Feurer R, Kihn Y, Vahlas C, Kalck P (2002) Parametric study for the growth of carbon nanotubes by catalytic chemical vapor deposition in a fluidized bed reactor. Carbon 40:1799–1807
49. Qin LC, Zhou D, Krauss AR, Gruen DM (1998) Growing carbon nanotubes by microwave plasma-enhanced chemical vapor deposition. Appl Phys Lett 72:3437–3439

50. Pirard SL, Pirard JP, Bossuot C (2009) Modeling of a continuous rotary reactor for carbon nanotube synthesis by catalytic chemical vapor deposition. AICHE J 55:675–686
51. Dupuis A (2005) The catalyst in the CCVD of carbon nanotubes—a review. Prog Mater Sci 50:929–961
52. Barzegar HR, Nitze F, Sharifi T, Ramstedt M, Tai CW, Malolepszy A, Stobinski L, Wagberg T (2012) Simple dip-coating process for the synthesis of small diameter single-walled carbon nanotubes—effect of catalyst composition and catalyst particle size on chirality and diameter. J Phys Chem C 116:12232–12239
53. Zhou K, Huang JQ, Zhang Q, Wei F (2010) Multi-directional growth of aligned carbon nanotubes over catalyst film prepared by atomic layer deposition. Nanoscale Res Lett 5:1555–1560
54. Wei YY, Eres G, Merkulov VI, Lowndes DH (2001) Effect of catalyst film thickness on carbon nanotube growth by selective area chemical vapor deposition. Appl Phys Lett 78:1394–1396
55. Gan B, Ahn J, Zhang Q, Rusli, Yoon SF, Yu J, Huang QF, Chew K, Ligatchev VA, Zhang XB, Li WZ (2001) Y-junction carbon nanotubes grown by in situ evaporated copper catalyst. Chem Phys Lett 333:23–28
56. Lee SY, Yamada M, Miyake M (2005) Synthesis of carbon nanotubes over gold nanoparticle supported catalysts. Carbon 43:2654–2663
57. Wong YM, Wei S, Kang WP, Davidson JL, Hofmeister W, Huang JH, Cui Y (2004) Carbon nanotubes field emission devices grown by thermal CVD with palladium as catalysts. Diam Relat Mater 13:2105–2112
58. Han JH, Choi SH, Lee TY, Yoo JB, Park CY, Jung T, Yu SG, Yi W, Han IT, Kim JM (2003) Growth characteristics of carbon nanotubes using platinum catalyst by plasma enhanced chemical vapor deposition. Diam Relat Mater 12:878–883
59. Blanco M, Alvarez P, Blanco C, Jimenez MV, Perez-Torrente JJ, Oro LA, Blasco J, Cuarterod V, Menendez R (2016) Enhancing the hydrogen transfer catalytic activity of hybrid carbon nanotube–based NHC–iridium catalysts by increasing the oxidation degree of the nanosupport. Cat Sci Technol 6:5504–5514
60. Fazil A, Chetty R (2014) Synthesis and evaluation of carbon nanotubes supported silver catalyst for alkaline fuel cell. Electroanalysis 26:2380–2387
61. Lee CJ, Lyu SC, Kim HW, Park JW, Jung HM, Park J (2002) Carbon nanotubes produced by tungsten-based catalyst using vapor phase deposition method. Chem Phys Lett 361:469–472
62. Zhou D, Wang S, Seraphin S (1994) Single-Walled Carbon Nanotubes Grown from Yttrium Carbide Particles. Fifty-Second Annual Meeting—Microscopy Society of America/Twenty-Ninth Annual Meeting—Microbeam Analysis Society, Proceedings, 772–773
63. Flahaut E, Peigney A, Bacsa WS, Bacsa RR, Laurent C (2004) CCVD synthesis of carbon nanotubes from (Mg, Co, Mo)O catalysts: influence of the proportions of cobalt and molybdenum. J Mater Chem 14:646–653
64. Botti S, Ciardi R, Terranova ML, Piccirillo S, Sessa V, Rossi M, Vittori-Antisari M (2002) Self-assembled carbon nanotubes grown without catalyst from nanosized carbon particles adsorbed on silicon. Appl Phys Lett 80:1441–1443
65. Azam MA, Isomura K, Ismail S, Mohamad N, Shimoda T (2015) Electrically conductive aluminum oxide thin film used as cobalt catalyst-support layer in vertically aligned carbon nanotube growth. Adv Nat Sci: Nanosci Nanotechnol 6
66. Chai SP, Zein SHS, Mohamed AR (2007) Synthesizing carbon nanotubes and carbon nanofibers over supported-nickel oxide catalysts via catalytic decomposition of methane. Diam Relat Mater 16:1656–1664
67. Cho W, Schulz M, Shanov V (2014) Growth and characterization of vertically aligned centimeter long CNT arrays. Carbon 72:264–273
68. Hata K, Futaba DN, Mizuno K, Namai T, Yumura M, Iijima S (2004) Water-assisted highly efficient synthesis of impurity-free single-walled carbon nanotubes. Science 306:1362–1364
69. Tessonnier JP, Su DS (2011) Recent progress on the growth mechanism of carbon nanotubes: a review. ChemSusChem 4:824–847

70. Kamachali RD (2006) Theoretical calculations on the catalytic growth of multiwall carbon nanotube in chemical vapor deposition. Chem Phys 327(2):434–438
71. Puretzky AA, Geohegan DB, Jesse S, Ivanov IN, Eres G (2005) In situ measurements and modeling of carbon nanotube array growth kinetics during chemical vapor deposition. Appl Phys A 81(2):223–240
72. Patole SP, Kim H, Choi J, Kim Y, Baik S, Yoo JB (2010) Kinetics of catalyst size dependent carbon nanotube growth by growth interruption studies. Appl Phys Lett 96(9):094101
73. Hasegawa K, Noda S (2011) Millimeter-tall single-walled carbon nanotubes rapidly grown with and without water. ACS Nano 5(2):975–984
74. Meshot ER, Hart AJ (2008) Abrupt self-termination of vertically aligned carbon nanotube growth. Appl Phys Lett 92(11):113107
75. Pal SK, Talapatra S, Kar S, Ci L, Vajtai R, Borca-Tasciuc T et al (2008) Time and temperature dependence of multi-walled carbon nanotube growth on Inconel 600. Nanotechnology 19(4):045610
76. Stadermann M, Sherlock SP, In JB, Fornasiero F, Park HG, Artyukhin AB, Wang YM, De Yoreo JJ, Grigoropoulos CP, Bakajin O, Chernov AA, Noy A (2009) Mechanism and kinetics of growth termination in controlled chemical vapor deposition growth of multiwall carbon nanotube arrays. Nano Lett 9:738–744
77. Christen HM, Puretzky AA, Cui H, Belay K, Fleming PH, Geohegan DB, Lowndes DH (2004) Rapid growth of long, vertically aligned carbon nanotubes through efficient catalyst optimization using metal film gradients. Nano Lett 4(10):1939–1942
78. Liu K, Jiang K, Wei Y, Ge S, Liu P, Fan S (2007) Controlled termination of the growth of vertically aligned carbon nanotube arrays. Adv Mater 19(7):975–978
79. Louchev OA, Laude T, Sato Y, Kanda H (2003) Diffusion-controlled kinetics of carbon nanotube forest growth by chemical vapor deposition. J Chem Phys 118(16):7622–7634
80. Mora E, Harutyunyan AR (2008) Study of single-walled carbon nanotubes growth via the catalyst lifetime. J Phys Chem C 112(13):4805–4812
81. Xiang R, Yang Z, Zhang Q, Luo G, Qian W, Wei F et al (2008) Growth deceleration of vertically aligned carbon nanotube arrays: catalyst deactivation or feedstock diffusion controlled? J Phys Chem C 112(13):4892–4896
82. Zhu L, Xu J, Xiao F, Jiang H, Hess DW, Wong CP (2007) The growth of carbon nanotube stacks in the kinetics-controlled regime. Carbon 45:344–348
83. Hofmann S, Csanyi G, Ferrari AC, Payne MC, Robertson J (2005) Surface diffusion: the low activation energy path for nanotube growth. Phys Rev Lett 95(3):036101
84. Pelech I, Narkiewicz U (2009) The kinetics of ethylene decomposition on iron catalyst. Acta Phys Pol A 116:S146–S149
85. Yamada T, Maigne A, Yudasaka M, Mizuno K, Futaba DN, Yumura M, Iijima S, Hata K (2008) Revealing the secret of water-assisted carbon nanotube synthesis by microscopic observation of the interaction of water on the catalysts. Nano Lett 8:4288–4292
86. Borjesson A, Bolton K (2011) Modeling of Ostwald ripening of metal clusters attached to carbon nanotubes. J Phys Chem C 115:24454–24462
87. Amama PB, Pint CL, Kim SM, Mcjilton L, Eyink KG, Stach EA, Hauge RH, Maruyama B (2010) Influence of alumina type on the evolution and activity of alumina-supported Fe catalysts in single-walled carbon nanotube carpet growth. ACS Nano 4:895–904
88. Louchev OA, Sato Y, Kanda H (2002) Growth mechanism of carbon nanotube forests by chemical vapor deposition. Appl Phys Lett 80:2752
89. Han JH, Graff RA, Welch B, Marsh CP, Franks R, Strano MS (2008) A mechanochemical model of growth termination in vertical carbon nanotube forests. ACS Nano 2:53–60
90. Blanch AJ, Lenehan CE, Quinton JS (2011) Parametric analysis of sonication and centrifugation variables for dispersion of single walled carbon nanotubes in aqueous solutions of sodium dodecylbenzene sulfonate. Carbon 49:5213–5228
91. Huang LP, Zhang HL, Wu B, Liu YQ, Wei DC, Chen JY, Xue YZ, Yu G, Kajiura H, Li YM (2010) A generalized method for evaluating the metallic-to-semiconducting ratio of

separated single-walled carbon nanotubes by UV-vis-NIR characterization. J Phys Chem C 114:12095–12098
92. Subbaiyan NK, Parra-Vasquez ANG, Cambre S, Cordoba MAS, Yalcin SE, Hamilton CE, Mack NH, Blackburn JL, Doorn SK, Duque JG (2015) Bench-top aqueous two-phase extraction of isolated individual single-walled carbon nanotubes. Nano Res 8:1755–1769
93. Espinosa HD, Filleter T, Naraghi M (2012) Multiscale experimental mechanics of hierarchical carbon-based materials. Adv Mater 24:2805–2823
94. Filleter T, Espinosa HD (2013) Multi-scale mechanical improvement produced in carbon nanotube fibers by irradiation cross-linking. Carbon 56:1–11
95. Zhao J, Shi DL, Lian J (2009) Small angle light scattering study of improved dispersion of carbon nanofibers in water by plasma treatment. Carbon 47:2329–2336
96. Chatterjee T, Jackson A, Krishnamoorti R (2008) Hierarchical structure of carbon nanotube networks. J Am Chem Soc 130:6934–6935
97. Mahdavi M, Baniassadi M, Baghani M, Dadmun M, Tehrani M (2015) 3D reconstruction of carbon nanotube networks from neutron scattering experiments. Nanotechnology 26:385704
98. Schaefer DW, Justice RS (2007) How nano are nanocomposites? Macromolecules 40:8501–8517
99. Jiang CM, Saha A, Young CC, Hashim DP, Ramirez CE, Ajayan PM, Pasquali M, Marti AA (2014) Macroscopic nanotube fibers spun from single-walled carbon nanotube polyelectrolytes. ACS Nano 8:9107–9112
100. Launois P, Marucci A, Vigolo B, Bernier P, Derre A, Poulin P (2001) Structural characterization of nanotube fibers by X-ray scattering. J Nanosci Nanotechnol 1:125–128
101. Kang MS, Shinb MK, Ismail YA, Shin SR, Kim SI, Kim H, Lee H, Kim SJ (2009) The fabrication of polyaniline/single-walled carbon nanotube fibers containing a highly-oriented filler. Nanotechnology 20:085701
102. Kozlov ME, Capps RC, Sampson WM, Ebron VH, Ferraris JP, Baughman RH (2005) Spinning solid and hollow polymer-free carbon nanotube fibers. Adv Mater 17:614–617
103. Zhou W, Vavro J, Guthy C, Winey KI, Fischer JE, Ericson LM, Ramesh S, Saini R, Davis VA, Kittrell C, Pasquali M, Hauge RH, Smalley RE (2004) Single wall carbon nanotube fibers extruded from super-acid suspensions: preferred orientation, electrical, and thermal transport. J Appl Phys 95:649–655
104. Huang W, Wang Y, Luo GH, Wei F (2003) 99.9% purity multi-walled carbon nanotubes by vacuum high-temperature annealing. Carbon 41:2585–2590
105. Luhrs CC, Garcia D, Tehrani M, Al-Haik M, Taha MR, Phillips J (2009) Generation of carbon nanofilaments on carbon fibers at 550 °C. Carbon 47:3071–3078
106. Tehrani M, Boroujeni AY, Luhrs C, Phillips J, Al-Haik MS (2014) Hybrid composites based on carbon fiber/carbon nanofilament reinforcement. Materials 7:4182–4195
107. Vaisman L, Wagner HD, Marom G (2006) The role of surfactants in dispersion of carbon nanotubes. Adv Colloid Interf Sci 128–130:37–46
108. Dehghan M, Al-Mahaidi R, Sbarski I, Mohammadzadeh E (2015) Surfactant-assisted dispersion of MWCNTs in epoxy resin used in CFRP strengthening systems. J Adhes 91:461–480
109. Xin F, Li L (2013) Effect of Triton X-100 on MWCNT/PP composites. J Thermoplast Compos Mater 26:227–242
110. Sahoo NG, Rana S, Cho JW, Li L, Chan SH (2010) Polymer nanocomposites based on functionalized carbon nanotubes. Prog Polym Sci 35:837–867
111. Breuer O, Sundararaj U (2004) Big returns from small fibers: a review on polymer/carbon nanotube composites. Polym Compos 25:630–645
112. Hussain F, Hojjati M, Okamoto M, Gorga RE (2006) Review article: polymer–matrix nanocomposites, processing, manufacturing, and application: an overview. J Compos Mater 40:1511–1575
113. Moniruzzaman M, Winey KI (2006) Polymer nanocomposites containing carbon nanotubes. Macromolecules 39:5194–5205

114. Manufacturing The Carbon Nanotube Market By John Evans, http://www.rsc.org/chemistryworld/Issues/2007/November/ManufacturingCarbonNanotubeMarket.asp Accessed May 2016. [Online]. [Accessed]. 2011. QSonica Sonicator Brochure. Cole Palmer
115. Montazeri A, Montazeri N, Pourshamsian K, Teharkhtchi A (2011) The effect of sonication time and dispersing medium on the mechanical properties of multiwalled carbon nanotube (MWCNT)/epoxy composite. Int J Polym Anal Charact 16:465–476
116. Sandler JKW, Kirk JE, Kinloch IA, Shaffer MSP, Windle AH (2003) Ultra-low electrical percolation threshold in carbon-nanotube-epoxy composites. Polymer 44:5893–5899
117. Han JH, Zhang H, Chen MJ, Wang GR, Zhang Z (2014) CNT buckypaper/thermoplastic polyurethane composites with enhanced stiffness, strength and toughness. Compos Sci Technol 103:63–71
118. OH JY, Yang SJ, Park JY, Kim T, Lee K, Kim YS, Han HN, Park CR (2015) Easy preparation of self-assembled high-density buckypaper with enhanced mechanical properties. Nano Lett 15:190–197
119. Kim SW, Kim T, Kim YS, Choi HS, Lim HJ, Yang SJ, Park CR (2012) Surface modifications for the effective dispersion of carbon nanotubes in solvents and polymers. Carbon 50:3–33
120. Xie XL, Mai YW, Zhou XP (2005) Dispersion and alignment of carbon nanotubes in polymer matrix: a review. Mater Sci Eng R-Rep 49:89–112
121. Yu J, Grossiord N, Koning CE, Loos J (2007) Controlling the dispersion of multi-wall carbon nanotubes in aqueous surfactant solution. Carbon 45:618–623
122. Tehrani M, Boroujeni AY, Hartman TB, Haugh TP, Case SW, Al-Haik MS (2013) Mechanical characterization and impact damage assessment of a woven carbon fiber reinforced carbon nanotube–epoxy composite. Compos Sci Technol 75:42–48
123. Tehrani M, Safdari M, Al-Haik MS (2011) Nanocharacterization of creep behavior of multiwall carbon nanotubes/epoxy nanocomposite. Int J Plast 27:887–901
124. Tehrani M, Safdari M, Boroujeni AY, Razavi Z, Case SW, Dahmen K, Garmestani H, Al-Haik MS (2013) Hybrid carbon fiber/carbon nanotube composites for structural damping applications. Nanotechnology 24, 155704
125. Zhang XF, Li QW, Holesinger TG, Arendt PN, Huang JY, Kirven PD, Clapp TG, Depaula RF, Liao XZ, Zhao YH, Zheng LX, Peterson DE, Zhu YT (2007) Ultrastrong, stiff, and lightweight carbon-nanotube fibers. Adv Mater 19:4198–4201
126. Behabtu N, Young CC, Tsentalovich DE, Kleinerman O, Wang X, Ma AW, Bengio EA, Ter Waarbeek RF, De Jong JJ, Hoogerwerf RE, Fairchild SB, Ferguson JB, Maruyama B, Kono J, Talmon Y, Cohen Y, Otto MJ, Pasquali M (2013) Strong, light, multifunctional fibers of carbon nanotubes with ultrahigh conductivity. Science 339:182–186
127. Zhao Y, Wei J, Vajtai R, Ajayan PM, Barrera EV (2011) Iodine doped carbon nanotube cables exceeding specific electrical conductivity of metals. Sci Rep 1:83
128. Di J, Wang X, Xing Y, Zhang Y, Zhang X, Lu W, Li Q, Zhu YT (2014) Dry-processable carbon nanotubes for functional devices and composites. Small 10:4606–4625
129. Lekawa-Raus A, Patmore J, Kurzepa L, Bulmer J, Koziol K (2014) Electrical properties of carbon nanotube based fibers and their future use in electrical wiring. Adv Funct Mater 24:3661–3682
130. Lee J, Stein IY, Devoe ME, Lewis DJ, Lachman N, Kessler SS, Buschhorn ST, Wardle BL (2015) Impact of carbon nanotube length on electron transport in aligned carbon nanotube networks. Appl Phys Lett 106:053110
131. Che YC, Wang C, Liu J, Liu BL, Lin X, Parker J, Beasley C, Wong HSP, Zhou CW (2012) Selective synthesis and device applications of semiconducting single-walled carbon nanotubes using isopropyl alcohol as feedstock. ACS Nano 6:7454–7462
132. Kim W, Choi HC, Shim M, Li YM, Wang DW, Dai HJ (2002) Synthesis of ultralong and high percentage of semiconducting single-walled carbon nanotubes. Nano Lett 2:703–708
133. Li JH, Ke CT, Liu KH, Li P, Liang SH, Finkelstein G, Wang F, Liu J (2014) Importance of diameter control on selective synthesis of semiconducting single-walled carbon nanotubes. ACS Nano 8:8564–8572

134. Li WS, Hou PX, Liu C, Sun DM, Yuan JT, Zhao SY, Yin LC, Cong HT, Cheng HM (2013) High-quality, highly concentrated semiconducting single-wall carbon nanotubes for use in field effect transistors and biosensors. ACS Nano 7:6831–6839
135. Liu BL, Liu J, Li HB, Bhola R, Jackson EA, Scott LT, Page A, Irle S, Morokuma K, Zhou CW (2015) Nearly exclusive growth of small diameter semiconducting single-wall carbon nanotubes from organic chemistry synthetic end-cap molecules. Nano Lett 15:586–595
136. Qian Y, Huang B, Gao FL, Wang CY, Ren GY (2010) Preferential growth of semiconducting single-walled carbon nanotubes on substrate by europium oxide. Nanoscale Res Lett 5:1578–1584
137. Qin XJ, Peng F, Yang F, He XH, Huang HX, Luo D, Yang J, Wang S, Liu HC, Peng LM, Li Y (2014) Growth of semiconducting single-walled carbon nanotubes by using ceria as catalyst supports. Nano Lett 14:512–517
138. Qu LT, Du F, Dai LM (2008) Preferential syntheses of semiconducting vertically aligned single-walled carbon nanotubes for direct use in FETs. Nano Lett 8:2682–2687
139. Loebick CZ, Podila R, Reppert J, Chudow J, Ren F, Haller GL, Rao AM, Pfefferle LD (2010) Selective synthesis of subnanometer diameter semiconducting single-walled carbon nanotubes. J Am Chem Soc 132:11125–11131
140. Song W, Jeon C, Kim YS, Kwon YT, Jung DS, Jang SW, Choi WC, Park JS, Saito R, Park CY (2010) Synthesis of bandgap-controlled semiconducting single-walled carbon nanotubes. ACS Nano 4:1012–1018
141. Komatsu N, Wang F (2010) A comprehensive review on seperation methods and techniques for single-walled carbon nanotubes. Materials 3:3818–3844
142. Subbaiyan NK, Cambre S, Parra-Vasquez ANG, Haroz EH, Doorn SK, Duque JG (2014) Role of surfactants and salt in aqueous two-phase separation of carbon nanotubes toward simple chirality isolation. ACS Nano 8:1619–1628
143. Chattopadhyay D, Galeska I, Papadimitrakopoulos F (2002) Complete elimination of metal catalysts from single wall carbon nanotubes. Carbon 40:985–988
144. Shim HC, Song JW, Kwak YK, Kim S, Han CS (2009) Preferential elimination of metallic single-walled carbon nanotubes using microwave irradiation. Nanotechnology 20:065707
145. Song JZ, Lu CF, Jin SH, Dunham SN, Xie X, Rogers JA, Huang YG (2015) Purification of single-walled carbon nanotubes based on thermocapillary flow. J Appl Mech 82:071010
146. Zhang GY, Qi PF, Wang XR, Lu YR, Li XL, Tu R, Bangsaruntip S, Mann D, Zhang L, Dai HJ (2006) Selective etching of metallic carbon nanotubes by gas-phase reaction. Science 314:974–977
147. Zhou WB, Fan QX, Zhang Q, Li KW, Cai L, Gu XG, Yang F, Zhang N, Xiao ZJ, Chen HL, Xiao SQ, Wang YC, Liu HP, Zhou WY, Xie SS (2016) Ultrahigh-power-factor carbon nanotubes and an ingenious strategy for thermoelectric performance evaluation. Small 12:3407–3414
148. Lalwani G, Gopalan A, D'agati M, Sankaran JS, Judex S, Qin YX, Sitharaman B (2015) Porous three-dimensional carbon nanotube scaffolds for tissue engineering. J Biomed Mater Res A 103:3212–3225
149. Stout DA (2015) Recent advancements in carbon nanofiber and carbon nanotube applications in drug delivery and tissue engineering. Curr Pharm Des 21:2037–2044
150. Stein J, Lenczowski B, Anglaret E, Frety N (2014) Influence of the concentration and nature of carbon nanotubes on the mechanical properties of AA5083 aluminium alloy matrix composites. Carbon 77:44–52
151. Chou T-W, Gao L, Thostenson ET, Zhang Z, Byun J-H (2010) An assessment of the science and technology of carbon nanotube–based fibers and composites. Compos Sci Technol 70:1–19
152. Coleman JN, Khan U, Blau WJ, Gun'ko YK (2006) Small but strong: A review of the mechanical properties of carbon nanotube–polymer composites. Carbon 44:1624–1652
153. Suhr J, Koratkar N, Keblinski P, Ajayan P (2005) Viscoelasticity in carbon nanotube composites. Nat Mater 4:134–137

154. Lima MD, Fang SL, Lepro X, Lewis C, Ovalle-Robles R, Carretero-Gonzalez J, Castillo-Martinez E, Kozlov ME, Oh JY, Rawat N, Haines CS, Haque MH, Aare V, Stoughton S, Zakhidov AA, Baughman RH (2011) Biscrolling nanotube sheets and functional guests into yarns. Science 331:51–55
155. Bakshi SR, Agarwal A (2011) An analysis of the factors affecting strengthening in carbon nanotube reinforced aluminum composites. Carbon 49:533–544
156. Kashiwagi T, Du FM, Douglas JF, Winey KI, Harris RH, Shields JR (2005) Nanoparticle networks reduce the flammability of polymer nanocomposites. Nat Mater 4:928–933
157. Beigbeder A, Degee P, Conlan SL, Mutton RJ, Clare AS, Pettitt ME, Callow ME, Callow JA, Dubois P (2008) Preparation and characterisation of silicone-based coatings filled with carbon nanotubes and natural sepiolite and their application as marine fouling-release coatings. Biofouling 24:291–302
158. Wu ZC, Chen ZH, Du X, Logan JM, Sippel J, Nikolou M, Kamaras K, Reynolds JR, Tanner DB, Hebard AF, Rinzler AG (2004) Transparent, conductive carbon nanotube films. Science 305:1273–1276
159. Cao Q, Rogers JA (2009) Ultrathin films of single-walled carbon nanotubes for electronics and sensors: a review of fundamental and applied aspects. Adv Mater 21:29–53
160. Ionescu AM, Riel H (2011) Tunnel field-effect transistors as energy-efficient electronic switches. Nature 479:329–337
161. Sun DM, Timmermans MY, Tian Y, Nasibulin AG, Kauppinen EI, Kishimoto S, Mizutani T, Ohno Y (2011) Flexible high-performance carbon nanotube integrated circuits. Nat Nanotechnol 6:156–161
162. Rueckes T, Kim K, Joselevich E, Tseng GY, Cheung CL, Lieber CM (2000) Carbon nanotube–based nonvolatile random access memory for molecular computing. Science 289:94–97
163. Kim P, Lieber CM (1999) Nanotube nanotweezers. Science 286:2148–2150
164. Dai LM, Chang DW, Baek JB, Lu W (2012) Carbon nanomaterials for advanced energy conversion and storage. Small 8:1130–1166
165. Kohler AR, Som C, Helland A, Gottschalk F (2008) Studying the potential release of carbon nanotubes throughout the application life cycle. J Clean Prod 16:927–937
166. An KH, Kim WS, Park YS, Moon JM, Bae DJ, Lim SC, Lee YS, Lee YH (2001) Electrochemical properties of high-power supercapacitors using single-walled carbon nanotube electrodes. Adv Funct Mater 11:387–392
167. Gong KP, Du F, Xia ZH, Durstock M, Dai LM (2009) Nitrogen-doped carbon nanotube arrays with high electrocatalytic activity for oxygen reduction. Science 323:760–764
168. Matsumoto T, Komatsu T, Arai K, Yamazaki T, Kijima M, Shimizu H, Takasawa Y, Nakamura J (2004) Reduction of Pt usage in fuel cell electrocatalysts with carbon nanotube electrodes. Chem Commun 7:840–841
169. Lee JM, Park JS, Lee SH, Kim H, Yoo S, Kim SO (2011) Selective electron- or hole-transport enhancement in bulk-heterojunction organic solar cells with N- or B-doped carbon nanotubes. Adv Mater 23:629–633
170. Gabor NM, Zhong ZH, Bosnick K, Park J, Mceuen PL (2009) Extremely efficient multiple electron-hole pair generation in carbon nanotube photodiodes. Science 325:1367–1371
171. Holt JK, Park HG, Wang YM, Stadermann M, Artyukhin AB, Grigoropoulos CP, Noy A, Bakajin O (2006) Fast mass transport through sub-2-nanometer carbon nanotubes. Science 312:1034–1037
172. Gao G, Vecitis CD (2011) Electrochemical carbon nanotube filter oxidative performance as a function of surface chemistry. Environ Sci Technol 45:9726–9734
173. Rahaman MS, Vecitis CD, Elimelech M (2012) Electrochemical carbon-nanotube filter performance toward virus removal and inactivation in the presence of natural organic matter. Environ Sci Technol 46:1556–1564
174. Corry B (2008) Designing carbon nanotube membranes for efficient water desalination. J Phys Chem B 112:1427–1434

175. Heller DA, Baik S, Eurell TE, Strano MS (2005) Single-walled carbon nanotube spectroscopy in live cells: Towards long-term labels and optical sensors. Adv Mater 17:2793–2799
176. Star A, Tu E, Niemann J, Gabriel JCP, Joiner CS, Valcke C (2006) Label-free detection of DNA hybridization using carbon nanotube network field-effect transistors. Proc Natl Acad Sci U S A 103:921–926
177. Liu Z, Tabakman S, Welsher K, Dai HJ (2009) Carbon nanotubes in biology and medicine: in vitro and in vivo detection, imaging and drug delivery. Nano Res 2:85–120
178. Kurkina T, Vlandas A, Ahmad A, Kern K, Balasubramanian K (2011) Label-free detection of few copies of DNA with carbon nanotube impedance biosensors. Angew Chem Int Ed Engl 50:3710–3714
179. Heller DA, Jin H, Martinez BM, Patel D, Miller BM, Yeung TK, Jena PV, Hobartner C, Ha T, Silverman SK, Strano MS (2009) Multimodal optical sensing and analyte specificity using single-walled carbon nanotubes. Nat Nanotechnol 4:114–120
180. De La Zerda A, Zavaleta C, Keren S, Vaithilingam S, Bodapati S, Liu Z, Levi J, Smith BR, Ma TJ, Oralkan O, Cheng Z, Chen XY, Dai HJ, Khuri-Yakub BT, Gambhir SS (2008) Carbon nanotubes as photoacoustic molecular imaging agents in living mice. Nat Nanotechnol 3:557–562
181. Esser B, Schnorr JM, Swager TM (2012) Selective detection of ethylene gas using carbon nanotube–based devices: utility in determination of fruit ripeness. Angew Chem Int Ed Engl 51:5752–5756
182. Snow ES, Perkins FK, Houser EJ, Badescu SC, Reinecke TL (2005) Chemical detection with a single-walled carbon nanotube capacitor. Science 307:1942–1945
183. Hong SY, Tobias G, Al-Jamal KT, Ballesteros B, Ali-Boucetta H, Lozano-Perez S, Nellist PD, Sim RB, Finucane C, Mather SJ, Green MLH, Kostarelos K, Davis BG (2010) Filled and glycosylated carbon nanotubes for in vivo radioemitter localization and imaging. Nat Mater 9:485–490
184. Bianco A, Kostarelos K, Prato M (2011) Making carbon nanotubes biocompatible and biodegradable. Chem Commun 47:10182–10188

Chapter 2
Synthesis, Characterization, and Applications of Carbon Nanotubes Functionalized with Magnetic Nanoparticles

Rakesh P. Sahu, Ahmed M. Abdalla, Abdel Rahman Abdel Fattah, Suvojit Ghosh, and Ishwar K. Puri

2.1 Introduction

Nanomaterials have a characteristic dimension of the order of 100 nm. They occur as compact materials or in dispersions. The deviation of their properties from those of bulk materials with the same chemical constituents has led to research across a wide range of applications. Various 1D nanomaterials include nanospheres, nanorods, nanobelts, nanorings, nanotubes, nanohelics, nanowires, and nanosheets, each with unique properties. Here, we discuss carbon nanotubes and magnetic nanoparticles and explain how their properties can be advantageously merged to overcome certain barriers and meet specific objectives.

R.P. Sahu • A.R.A. Fattah
Department of Mechanical Engineering, McMaster University,
1280 Main Street West, Hamilton, ON, Canada, L8S 4L7
e-mail: sahur@mcmaster.ca; abdelfar@mcmaster.ca

A.M. Abdalla • S. Ghosh
Department of Engineering Physics, McMaster University,
1280 Main Street West, Hamilton, ON, Canada, L8S 4L7
e-mail: abdallam@mcmaster.ca; sghosh@mcmaster.ca

I.K. Puri (✉)
Department of Mechanical Engineering, McMaster University,
1280 Main Street West, Hamilton, ON, Canada, L8S 4L7

Department of Engineering Physics, McMaster University,
1280 Main Street West, Hamilton, ON, Canada, L8S 4L7
e-mail: ikpuri@mcmaster.ca

G. Balasubramanian (ed.), *Advances in Nanomaterials*,
DOI 10.1007/978-3-319-64717-3_2

2.1.1 Carbon Nanotubes

Carbon nanotubes are anisotropic 1-D structures. After multi-walled carbon nanotubes were discovered using high-resolution transmission electron microscopy [1, 2], research revealed the structure and properties of other types of CNTs, such as single-walled nanotubes (SWNTs) and double-walled nanotubes (DWNTs) along with carbonaceous nanomaterials like graphene. SWNTs consist of a single 1D graphene sheet rolled into a tube with a diameter of 1–2 nm and a relatively high aspect ratio [2]. MWNTs consist of multiple graphene sheets rolled into cylinders of outer diameter 10–80 nm [3]. The structure of CNTs is revealed through the chiral vector **c** (expressed by two indices n and m) and the chiral angle θ. Figure 2.1 illustrates the three broad CNT types that depend on the rolling orientation of the chiral vector, namely (a) armchair ($n = m$), (b) zigzag ($m = 0$), and (c) chiral ($n \geq 0$: $m \geq 0$) nanotubes, where $0° < \theta < 30°$, i.e., θ is the chirality angle. The unique electrical properties of a CNT depend on its chirality and tube diameter. CNTs can be either metallic (armchair nanotubes) when $|n - m| = 3q$ where q is an integer, or semiconducting for all other cases. The extraordinary electrical properties of SWNTs have pointed the way for fabricating novel electronics [4]. The shape and structure of a CNT results in a high axial thermal conductivity of 1750–5800 W m^{-1} K^{-1} at room temperature [5]. The electrical resistivity of CNTs has been measured in the range 10^{-4}–10^{-3} Ω cm along with an exceedingly high current carrying capacity up to 10^9 A cm^{-2} [5–7]. CNTs have high thermal stability (up to 2800 °C in vacuum and about 750 °C in air), high surface area (200–900 m^2 g^{-1}), low density (1–2 g cm^{-3}), and high Young's modulus (1–1.8 TPa) [3, 5, 6]. These properties make CNTs useful for sensors [8, 9], high strength materials [10], nanoelectronics [11, 12], fuel storage [13], energy storage [14], and biomedicine [15–18].

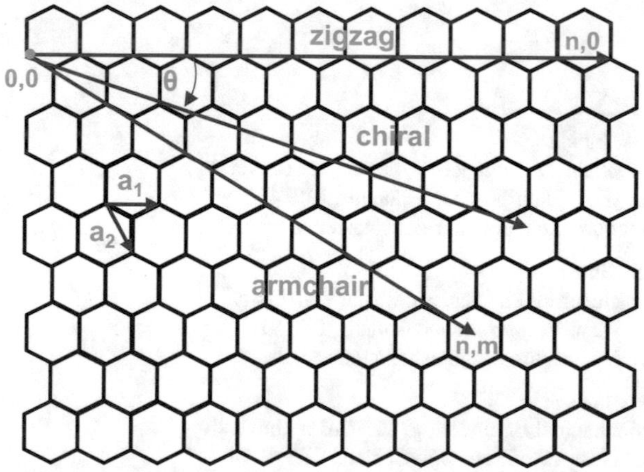

$$\vec{c} = n\vec{a_1} + m\vec{a_2} = (n,m)$$

Fig. 2.1 Schematic of the graphene sheet and the different structures of a nanotube that can be rolled based on the chiral vector **c** characterized by the indices n and m

Commercially available CNTs are typically produced by chemical vapor deposition [19–21], and are entangled and randomly oriented due to short-range van der Waals forces between adjacent CNTs. This entanglement diminishes the effective properties of CNTs [22–25]. Being nonpolar, the carbon atoms of CNTs have a high affinity toward nonpolar materials, such as organic solvents, oils, and hydrocarbons. CNT's hydrophobicity limits their dispersion stability with polar solvents and polymers [26, 27]. Pure CNTs are diamagnetic but those synthesized with metal nanoparticles as catalysts can exhibit ferromagnetic behavior due to the presence of the magnetic catalyst [28], which is easily lost during acid treatment.

The diamagnetic susceptibility of CNTs limits remote control, or action from a distance, using a magnetic field. The chemical inertness of CNTs poses a serious difficulty for synthesizing composites with materials that are important for device applications. Hence, CNT functionalization with MNPs is a strategy to chaperone the nanotubes, manipulate, and organize them on demand [29–33].

2.1.2 *Magnetic Nanoparticles*

The magnetic properties of MNPs are used in many applications, e.g., microwave absorption [34], electrochemical sensing [35], ferrofluids [36], energy storage [37], magnetic resonance imaging, and data storage [38]. Since these magnetic properties are dependent on MNP size and shape, as shown in Fig. 2.2, different synthesis methods have been explored, including microemulsion [39], thermal decomposition [40], co-precipitation [41], sol-gel [42], wet chemical [43], self-assembly [44], spray pyrolysis [45], solvothermal method [46], template directed [47], and

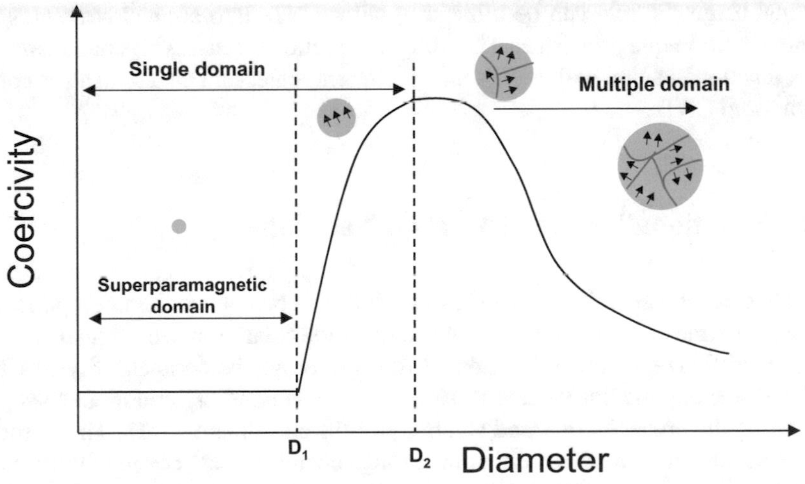

Fig. 2.2 Schematic representation of the change in coercivity with the size of a magnetic nanoparticle

deposition [48]. Although the physical properties of MNPs are improved with a large surface area to volume ratio, their agglomeration is an important concern. The crystallinity and the degree of defects or impurities of MNPs depend on the method of synthesis, which influences their magnetic behavior [49].

The five basic types of magnetism are ferromagnetism, antiferromagnetism, ferrimagnetism, paramagnetism, and diamagnetism. Ferrimagnetism occurs for compound materials, such as ferrites whereas the other types arise in pure elements. The spinning of electrons creates magnetic moments. Ferromagnetic materials have aligned magnetic moments that offer spontaneous magnetization in materials such as Fe, Ni, and Co. In antiferromagnetic materials, the magnetic moments are arranged in an antiparallel fashion so that the net magnetic moment is zero, a behavior that is observed at low temperatures. Ferrimagnets retain their magnetization even in the absence of a field but have antiparallel magnetic moments similar to antiferromagnets that are of unequal magnitude. Materials with uncoupled magnetic moments display paramagnetism with a small positive magnetic susceptibility. Materials that are repelled by a magnetic field and have a slightly negative susceptibility display diamagnetism.

The Weiss, or magnetic, domain is a volume of magnetic material in which all of the magnetic moments are aligned in the same direction. As the size of a ferromagnetic nanoparticle decreases, its magnetization becomes more uniform until a dimension D_2 when the domain walls within the particle disappear, resulting in a single domain. Further size reduction below D_1 results in thermal fluctuations overcoming the magnetic moment in the single domain so that the particle becomes superparamagnetic. Conversely, with an increase in particle size, the number of domain boundaries within a particle increases and thus its coercivity also decreases.

The repeatable synthesis of MNPs with a particular morphology and size is of consequence for monodisperse colloids. The magnetic response of MNPs to an external magnetic field can be utilized to tailor CNTs into aligned structures that harness their unique properties. The following section discusses recent advances of CNT functionalization with MNPs using different routes and the applications of the magnetized CNTs.

2.2 Functionalization of Carbon Nanotubes

The physical and chemical properties of CNTs can be enhanced in comparison to those of pristine CNTs by introducing external molecular groups or radicals on their surfaces [50]. These added functional groups improve the compatibility of CNTs with a dispersing medium and with solvents, polymer and organic molecules [51], improving the processability and solubility of the constituent CNTs. The introduction of a molecular group can either be through chemical (e.g., covalent interaction) or physical (e.g., van der Walls interaction, adsorption) means. Different CNT functionalization methods are classified based on the chemistry involved and illustrated in Fig. 2.3.

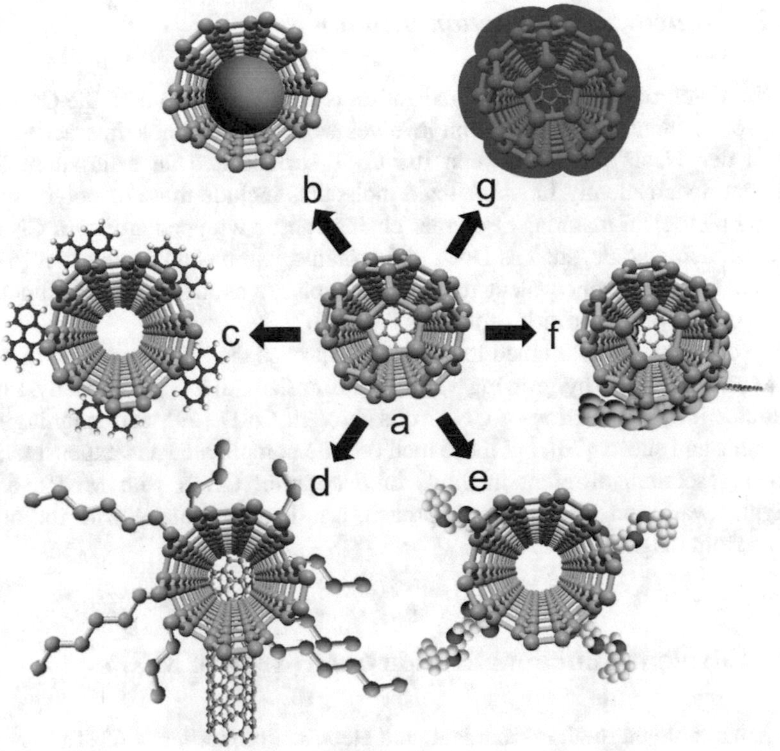

Fig. 2.3 Schematic of different methods of functionalizing SWNTs: (**a**) Single-walled carbon nanotube, (**b**) endohedral functionalization with, for example, C60, (**c**) covalent sidewall functionalization, (**d**) defect-group functionalization, (**e**) noncovalent exohedral functionalization with surfactants, (**f**) noncovalent exohedral functionalization with polymers, and (**g**) metal plating of carbon nanotubes

2.2.1 Covalent Functionalization

Several routes used to covalently functionalize CNTs include hydrogenation [52], electrochemical functionalization [53], thiolation [54], oxidative purification [55], halogenation [56], esterification [57], amidation [58], and cycloaddition [59]. Covalent functionalization allows functional groups to form covalent linkages on the carbon scaffold of nanotubes. CNTs inevitably contain defect sites during their production, which make the nanotubes susceptible to attack by reactive molecular groups. Acid treatment transforms defect sites into active molecular groups, such as COOH, C-OH, and C=O [51, 60] that are covalently attached to the CNT wall. Metal or metal oxide nanoparticles, charged polymer chains, and charged molecular groups can be covalently attached to these active sites, producing a strong bond in comparison to noncovalent functionalization, but at the expense of significant surface destruction.

2.2.2 *Noncovalent Functionalization*

The drawback of covalent functionalization is the damage done to the CNT structure. Noncovalent functionalization involves weak dipole–dipole interactions, such as van der Waals forces between the CNT surface and an external molecule. Different noncovalently functionalized molecules include those of polymers, metals, and biological materials. Polymer chains can be wrapped around a CNT wall [61] and biomaterials, such as DNA and proteins, can be conjugated to it [62–66]. Both covalent and noncovalent functionalization are exohedral, where functionalization occurs on the outside of the CNT wall [67].

The organization of aligned long CNTs improves the bulk properties of a composite material, e.g., by growing vertical and horizontal CNTs on catalyst coated substrates [68], production of CNT rolls through CVD [69], and spinning CNTs into yarns and sheets [70], but these methods are complicated and expensive. In the following section, different methods of decorating CNTs with MNPs, a more straightforward and inexpensive approach, are discussed along with the benefits arising from magnetized CNTs.

2.3 Covalent Functionalization of CNTs with MNPs

The remarkable thermal, mechanical, and electrical properties of CNTs have made them a promising material for a wide range of applications, including metal matrix composites, nanosensors, and reinforced polymer composites. These properties can be leveraged to form composite materials by embedding them into a polymer matrix [22, 71], particularly by aligning them in a particular direction [24]. CNTs are diamagnetic with a diamagnetic susceptibility χ of 10^{-5} emu g^{-1} [72, 73], making them resistant to manipulation with a magnetic field. However, magnetizing CNTs with superparamagnetic nanoparticles enables such a response, e.g., by intercalating them within CNTs [74–77], and synthesizing MNPs on the walls of CNTs [78–80]. The synthesis of ferritic nanoparticles is a convenient method of magnetizing CNTs since it requires inexpensive reagents and uses common laboratory apparatus. Ferritic nanoparticles can be synthesized on CNTs using several routes, such as surface imprinting of magnetite (Fe_3O_4) crystals [81], hydrothermal decomposition of iron compounds [82, 83], and in situ co-precipitation of ferrite crystals [84–86]. Surface imprinting requires a high-temperature autoclave whereas hydrothermal decomposition requires both a high temperature (~250 °C) and an inert environment. Co-precipitation is thus the most convenient method among these for decorating CNTs with ferritic nanoparticles.

2.3.1 *Co-precipitation of MNPs: Methodology*

The co-precipitation of MNPs on CNTs is carried through two steps. CNTs are initially treated with strong oxidizing acids such as HNO_3 and H_2SO_4. The NO_3^- and HSO_4^- radicals attack the C–C bonds at defect sites on the CNT surfaces that are formed during their production, forming functional groups, e.g., C=O, C-OH, and COOH. These active groups act as nucleation sites for the magnetite nanocrystals that are co-precipitated from a solution containing Fe^{2+} and Fe^{3+} ions [78, 87, 88].

A detailed and comparative study of decorating CNTs with magnetite nanoparticles using HNO_3, H_2SO_4, and a 1:1 mixture of HNO_3 and H_2SO_4 has been performed to determine the best practice to functionalize CNTs with minimum weight loss and higher magnetization density [89]. The effects of the intermediate stage of filtration and washing of the acid-treated CNTs are also investigated to understand its influence on the yield of the magnetized CNTs. Acid treatment of CNTs increases the oxygen content of the nanotubes. XPS analysis of different CNT samples with identical masses that have been treated with different acids is shown in Fig. 2.4. The increase in the oxygen content when CNTs are treated with a mixture of both acids is significantly higher than when individual acids are used for treatment, indicating severe structural damage to the CNTs, which consequently reduces their lengths. At the intermediate stage of washing, these shortened CNTs are washed away and lost to the filtrate. Attempts to skip the stage of washing during magnetizing CNTs resulted in acid residues in the sample which reacts with the iron ions in the solution.

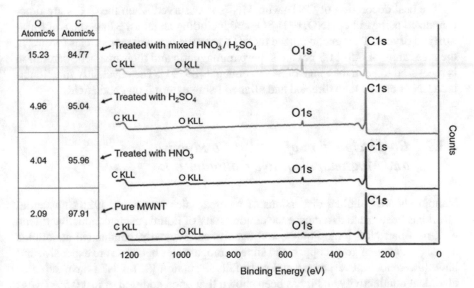

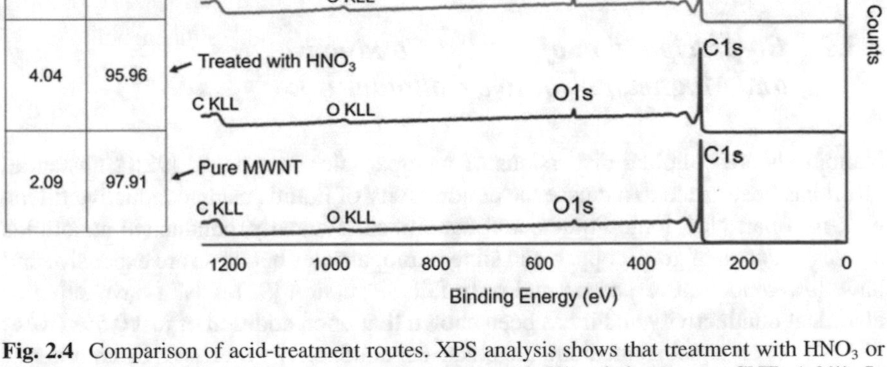

Fig. 2.4 Comparison of acid-treatment routes. XPS analysis shows that treatment with HNO_3 or H_2SO_4 leads to a moderate increase in oxygen content (4–5%) relative to pure CNTs (~2%). In contrast, treatment with a 1:1 mixture of the two acids produces a much higher (15.23%) oxygen content. Reprinted from [89] with permission from Elsevier

Table. 2.1 Yields of the various material phases, as a fraction of their stoichiometrically designed values

Material phase	HNO_3	H_2SO_4	(HNO_3:H_2SO_4) 1:1 (v/v)	HNO_3	H_2SO_4	(HNO_3:H_2SO_4) 1:1 (v/v)
	With the washing and drying stage			Without the washing and drying stage		
CNTs	90.2%	91.6%	54%	100%	78%	82%
Fe_3O_4	100%	94%	85.2%	76.8%	31%	60%
mCNTs (total yield)	95.3%	92.8%	69.6%	88.4%	54.5%	71%
Fe_3O_4:CNTs (w/w)	1.11	1.03	1.58	0.77	0.39	0.73

The magnetization ratio γ is the ratio of Fe_3O_4:CNTs (w/w). The intended value of this ratio is unity for all samples. Reprinted from [89] with permission from Elsevier

As a consequence, undesired materials are found to be co-precipitated such as Akaganeite (Fe^{3+} O(OH, Cl), PDF No. 00-013-0157), magnetic iron oxide hematite (α-Fe_2O_3, PDF No. 00-032-0469), ferrous sulfate ($FeSO_4$, PDF No. 00-042-0229) which compromised the magnetization ratio (MNP:CNT) of CNT. Table 2.1 lists the yield of CNTs, ferrite MNPs, magnetized CNTs, and the final magnetization weight ratio (γ) at different stages of decorating CNTs with MNPs that includes the acid treatment, washing, and drying. The filtration and drying step was found to be necessary after the acid treatment to avoid the co-precipitation of unnecessary non-magnetic products.

The best decoration of CNTs with MNPs is observed when the CNTs are functionalized using either HNO_3 or H_2SO_4 and including an intermediate stage of filtration and drying. The functionalized mCNTs with the ferrite nanoparticles formed in the size range of 8.5–11.3 nm are superparamagnetic and their magnetic saturation is measured to be in the range of 34–38 emu g^{-1} [89–91]. The covalently functionalized CNTs can be then directed and aligned using external magnetic field.

2.3.2 *Co-precipitation of MNPs: Conductive and Magnetoresponsive Colloidal Ink*

Nanofluids are colloidal dispersions of nanoparticles in a liquid [92]. Substantial effort has been made to increase the conductivity of liquids using conductive fillers, e.g., nanoparticles, ionic liquids, and nanotubes. Typically, conductive nanofluids are suspensions of gold, copper, and silver nanoparticles but these are expensive and have low oxidation resistance and only fair dispersion [93]. CNTs have superior electrical conductivity and it has been shown that upon addition of just 0.5% (w/w) of CNTs to an aqueous medium, the electrical conductivity increases by an order of magnitude [94]. Use of CNTs and graphene to prepare conductive colloidal suspensions has been reported but their agglomeration due to inter-CNT attraction poses a serious problem [95–98]. Temporary dispersions of CNTs are achieved by ultrasonication [99] but CNTs need to be functionalized with surfactant molecules in order to achieve long-term stability [100]. Ferrofluids are a type of nanofluids consisting of MNPs that respond to a magnetic field [29, 30, 32]. MNPs coated with surfactant produce interparticle repulsion thus increasing their colloidal stability [97, 101]. Ferrofluids that can be manipulated using a magnetic field typically have poor electrical conductivity whereas CNT dispersions provide much higher electrical conductivity. A method to combine the properties of both CNTs and MNPs allows the magnetic manipulation of conductive colloids that is useful for printing electronic circuits and sensors.

Magnetoresponsive conductive colloids are synthesized by co-precipitating MNPs on CNTs and dispersing them in aqueous phase [102]. The material content of the MNPs and the magnetization weight ratio are varied to study their influence on both of the electrical and magnetic properties of the dispersion. The MNPs are placed on CNTs by co-precipitation [89]. Three different MNPs that have co-precipitated on the CNTs for the magnetization weight ratio of unity are magnetite (S1), Mn-Cu-Zn ferrite (S2), and Cu-Zn Ferrite (S3). These MNPs are attached to the outside surface of CNTs via covalent bonds. The nanoparticles co-precipitated on CNTs, of 10–15 nm size, successfully decorate CNTs as is evident from Fig. 2.5. The XRD analysis shows that all the samples are crystalline with spinel structure. As the magnetization ratio increases, the decoration density of MNPs on CNTs increases with the nanoparticles occupying the available activation sites until a certain value when all the sites are occupied. The excess metal ions are then co-precipitated over the deposited layer of MNPs perpendicular to the axis of CNTs.

Magnetite has a high Curie temperature whereas Mn Cu Zn ferrite and Cu Zn ferrite have lower Curie temperatures [103–105]. The magnetic property of the colloidal dispersion containing magnetic CNTs (mCNTs) can thus be tailored by changing the material of the co-precipitated MNPs. The hysteresis curve for the three different mCNTs shows a complete absence of coercive field (Fig. 2.6a), indicating that the nanoparticles are superparamagnetic at the room temperature. The magnetic saturations M_s of the three specimens are measured to be 30.7, 10.5, and 16.6 emu g^{-1} for S1, S2, and S3 respectively. The sensitivities of each specimen to varying temperature are different for each sample, as shown in Fig. 2.6b. This demonstrates how the magnetic properties of functionalized CNTs can be tuned using different nanoparticles.

Tetramethylammonium hydroxide (TMAH: 25%) is used to suspend the functionalized CNTs in the aqueous phase. TMAH, on one hand, increases the electrical conductivity of the solution and, on the other hand, it acts as an ionic surfactant to stabilize the functionalized CNTs in a colloidal suspension. A schematic of the proposed rearrangement of the ions under the application of an electric field is shown in Fig. 2.6c. This polarization shortens the ion transport path and thus reduces the electrical resistance of the suspension. The addition of 4 wt% of the mCNTs results in increased electrical conductivity of the colloidal suspension (cf. Fig. 2.6d), and the increase is different for different nanoparticles. This methodology of preparing magnetoresponsive and electrically conducting colloidal dispersions is useful for applications where a magnetic response must be coupled with the material electrical conductivity.

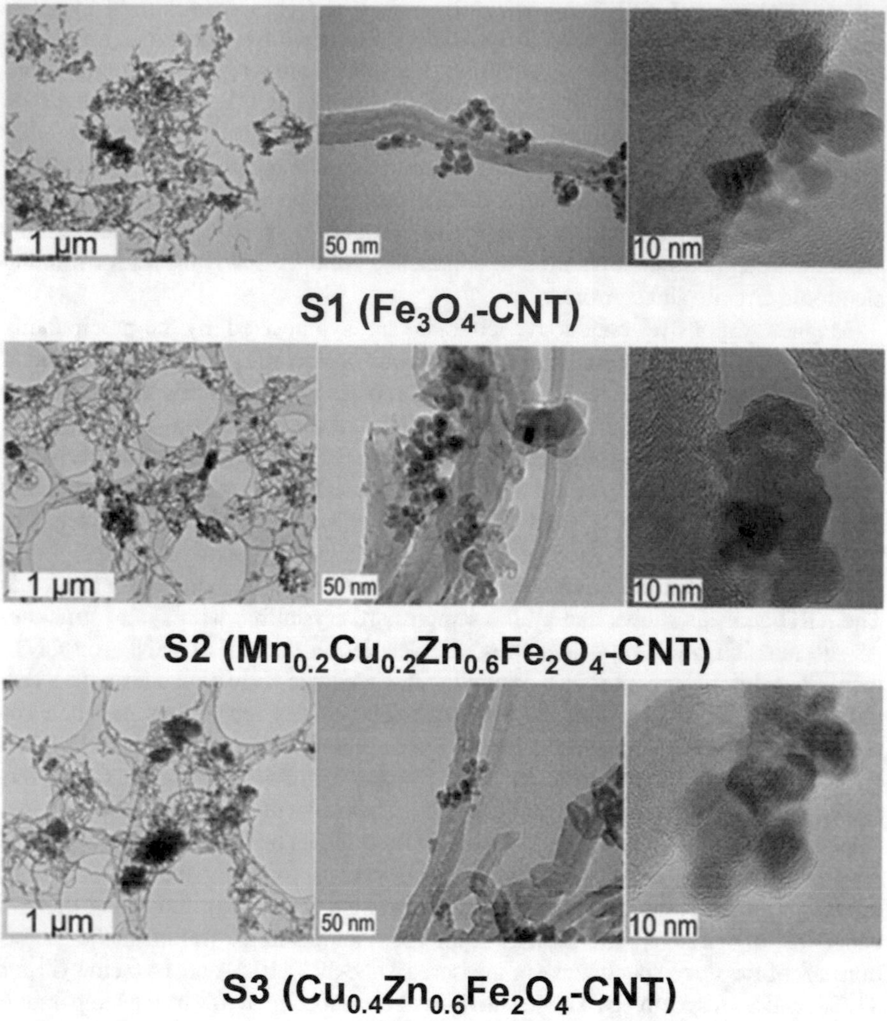

Fig. 2.5 TEM images of samples S1–S3 (from top to bottom) confirm that all samples corresponding to $\gamma = 1$ are successfully decorated with highly crystalline (but different) MNPs, synthesized within the narrow size distribution of 10–15 nm. Reprinted from [102] with permission from Elsevier

2.4 Noncovalent Functionalization of CNTs with MNPs

2.4.1 Metal Plating of CNT

Functionalization of CNTs with MNPs allows their manipulation and alignment using a magnetic field. The electrical conductivity of the aligned CNTs still depends on the interfacial contacts between adjacent nanotubes. Researchers have

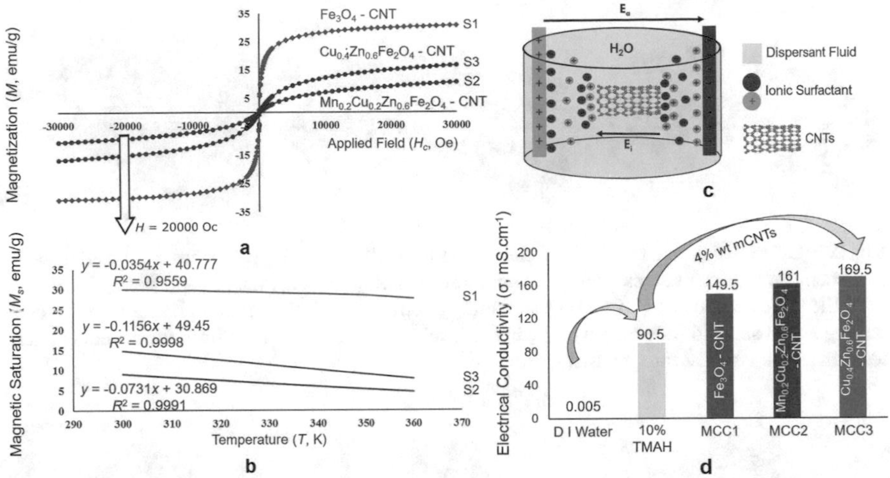

Fig. 2.6 (**a**) Magnetic hysteresis curves of the dry powders show no evidence of remanence, i.e., the mCNTs are superparamagnetic. (**b**) For all samples, M_s decreases with increasing temperature. Magnetite (S1) has the strongest magnetization and weakest sensitivity to temperature, Mn–Cu–Zn ferrite (S2) has the weakest magnetization while Cu–Zn ferrite (S3) has the strongest temperature sensitivity. (**c**) CNTs placed in an ionic medium between two electrodes charged by an electric field E_a polarizes and become oriented along the direction of the field. (**d**) Dissolving 10% (w/w) of tetramethyl ammonium hydroxide (TMAH) in DI water increases the electrical conductivity to 90.5 mS cm^{-1}. Dispersing 4% (w/w) of the different mCNTs in the TMAH solution enhances the electrical conductivity by 65–90%. Reprinted from [102] with permission from Elsevier

demonstrated deposition of metals or metal-based compounds over the nanotubes to decrease the contact resistance and also improve their resistance to oxidation [106]. Electroless plating of CNTs with metals like Ni, Cu, or Co can be performed under ambient conditions to produce uniform metal coatings. Electroless plating is a non-destructive and rapid method to deposit metal, which is initiated by activation-sensitization process that introduces catalytic nuclei on the surfaces of CNTs. A reduction reaction for the metal ions on this new catalytic surface forms a coating layer composed of metal. The metal-coated CNTs thus possess superior hardness, wear resistance, high electrical conductivity, and magnetic properties. A continuous and uniform layer of nickel can be electroplated on the surface of CNTs in a water bath without sensitizing or activating the surface with catalysts [107].

We have provided a controlled process for plating nickel over CNTs using electroless plating for different weight ratios of Ni:CNT (γ_e) [108], the schematic of which is shown in Fig. 2.7. Sensitization of CNTs using tin ions followed by activation with palladium ions reduces the nickel ions to nickel, which is deposited on the CNT outer surfaces. Nickel then acts as auto-catalyst that continues to reduce nickel ions until their complete depletion from the solution. This allows us to control the thickness of the deposited nickel layer over the MWNTs and thus their material properties.

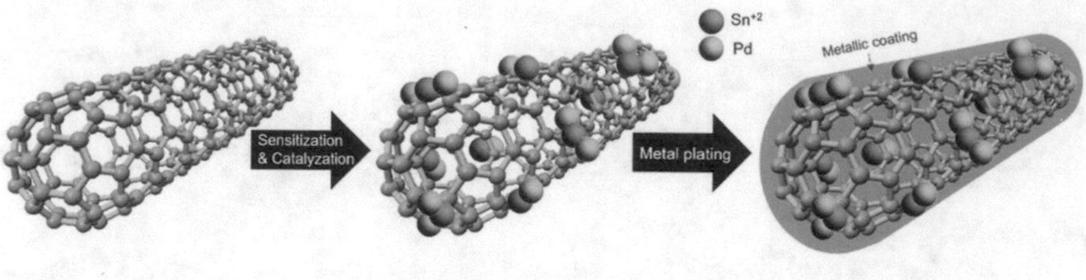

Fig. 2.7 CNTs magnetized with Ni by electroless deposition. (**a**) MWNTs were catalyzed through two chemical treatment steps using acid solutions of $SnCl_2$ for sensitization and $PdCl_2$ for activation. Electroless deposition of Ni on the resulting activated MWNTs used a plating solution containing nickel salt and reducing agent, where nickel ions accept electrons from the reducing agent to form metallic nickel through metal reduction

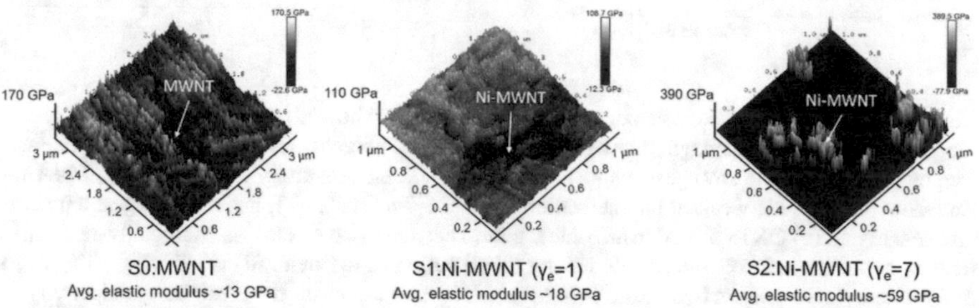

Fig. 2.8 AFM mechanical sketch up of samples S0, SN1, and SN2. Pure MWNTs (S0) had an average elastic modulus $E{\sim}13$ GPa while Ni-MWNT samples for $\gamma_e = 1$ and 7 had values of $E{\sim}18$ (46% increase) and 59 GPa (370% increase), respectively. Increasing Ni:MWNT weight ratio γ_e enhances the measured modulus

The morphology and thickness of the deposited layer of nickel and its mechanical and magnetic properties depends on the Ni:CNT weight ratio in the reacting mixture. The magnetic property of the hybrid nanomaterial increases with nickel weight fraction. As γ_e increases from 1 to 7, the magnetic saturation (M_s) and remnant magnetization (M_r) increase from 4.1 and 0.51 emu g^{-1} to 9.5 and 1.01 emu g^{-1} respectively. The average elastic modulus E of the hybrid CNTs ($\gamma_e = 7$) measured using AFM in the radial direction shows a threefold increase over that for pristine CNTs. Figure 2.8 shows the effect of nickel deposition on the elastic modulus of the CNTs, providing a solution to control it. The increase in elastic modulus is attributed to the thickness of the nickel deposit over the CNT surfaces, which increases with γ_e.

2.4.2 *Physical Attachment of MNPs on CNT*

CNTs can be decorated by ferrites nanoparticles, which improve the optical, magnetic, and electrochemical properties of pristine CNTs. The vapor deposition of Ni atoms allows decoration of nickel nanoparticles on the CNT surfaces. The process can be continued longer to coat the CNT with a layer of Ni [109]. The CNT surfaces can be sensitized and catalyzed to further reduce the metal ions to form nanoparticles. Both metal plating and physical attachment of MNPs onto the surface of CNTs require several steps that are time-consuming and laborious. Each method of functionalization has its advantages but CNT manipulation with a magnetic field is the objective, a simple one-step method is more desirable in comparison to a longer chemical route.

2.4.3 Entanglement of NiNP in CNT Network

Covalent functionalization of MNPs on CNTs often requires expensive chemicals and an experienced skillset. Entanglement of nickel nanoparticles (NiNP) in the matrix of CNTs using probe sonication is another novel method to create a magnetic ink [110]. Sonication of a dispersion of NiNPs and CNTs in kerosene (Fig. 2.9a) allows nickel to become entangled within the CNT network as shown in the TEM image (Fig. 2.9b). The high surface energy of NiNPs and pi-interactions are responsible for the clustering of NiNPs and their conjugation with CNTs. The conjugated NiNP-CNT material responds as a bulk to a magnetic field, which is evident from Fig. 2.9a. The conjugated material shows moderate magnetization saturation value of 14.61 emu g^{-1} (Fig. 2.9c). The nanoparticles are individually superparamagnetic but with a coercive field the material becomes ferromagnetic due to the entanglement of the nanoparticles responsible for inter-NiNP magnetostatic interactions for particles in close proximity. Prolonged sonication in most cases does not result in more homogeneously distributed NiNPs and thus is not necessary. Therefore, adequate magnetization usually occurs within a few minutes. For this reason, mechanical magnetization allows for rapid entanglement of NiNPs and offers one of the quickest methods to magnetize CNTs while forgoing harsh chemical pre-treatments familiar to other magnetization routes.

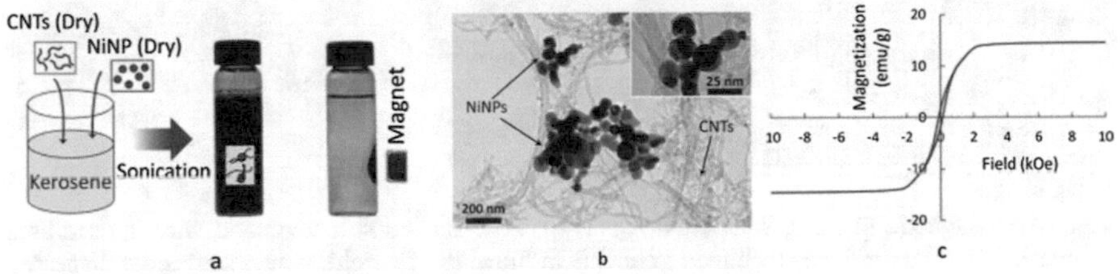

Fig. 2.9 Nickel nanoparticle entangled carbon nanotubes. (**a**) NiNPs and CNTs are dispersed in kerosene by probe sonication. The CNTs entangle the NiNPs, enabling the latter to act as magnetic chaperones without detachment or separation in a strong gradient magnetic field. (**b**) TEM images show clusters of NiNPs enmeshed in CNTs. (**c**) SQUID magnetometry shows that the NiNP-CNT, containing 33% (w/w) NiNPs, have a saturation magnetization M_s = 14.61 emu g^{-1}, which is commensurate with the mass fraction of nickel in the sample. Reprinted (adapted) with permission from [110]. Copyright (2016) American Chemical Society

2.4.4 Printing Sensors with Magnetized CNT Ink

We printed the synthesized magnetic ink into U shapes over polydimethylsiloxane (PDMS) surface using an iron template and a permanent magnet. The iron template placed on the permanent magnet concentrates the magnetic lines of force, producing a high gradient magnetic field. This field concentrates the NiNP-CNTs and aligns them along the magnetic lines of force (cf. Fig. 2.10a). On drying of the solvent in the magnetic ink, the NiNP-CNTs are embedded in the PDMS matrix (cf. Fig. 2.10b). The embedded U-shaped NiNP-CNT network on PDMS forms a continuous electrically conductive path, which we tested as an alternative to flexible sensors. It is an alternative to rigid conducting wires and appropriate for wearable electronics [111]. An easy, inexpensive, and facile fabrication method for flexible sensors using NiNP-CNT networks is used to fabricate strain gauge [112] and detect chemical and biological species [113, 114].

A simple voltage divider circuit is used to determine the response of the U-shaped NiNP-CNT network subjected to mechanical deformation, including both bending and elongation. The resistance of the NiNP-CNT circuit increases as the space between the interconnects increases and the percolation of the CNT network

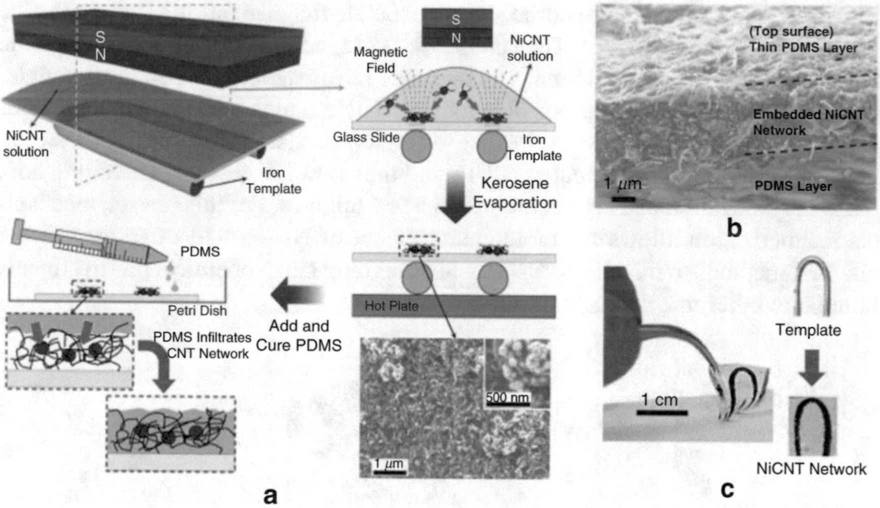

Fig. 2.10 Magnetic printing with NiNP-CNT: (**a**) A U-shaped soft magnetic wire is used as a template. The wire produces localized gradients in the magnetic field, which settles the dispersed NiNP-CNT on the coverslip surface immediately adjacent to it. The kerosene is then evaporated by heating, yielding a U-shaped dense network of CNTs. The NiNP-CNT U network is placed inside a Petri dish, covered with liquid PDMS, and heated to cure the PDMS. The PDMS infiltrates the printed structure, and after curing, it lends a polymer matrix to the NiNP-CNT. (**b**) SEM image of a cross-section of such a structure reveals a ~2.5 μm thick NiNP-CNT -PDMS composite network embedded in pure PDMS. (**c**) When the PDMS is peeled off, it forms a flexible and stretchable membrane with an embedded NiNP-CNT structure. Reprinted (adapted) with permission from [110]. Copyright (2016) American Chemical Society

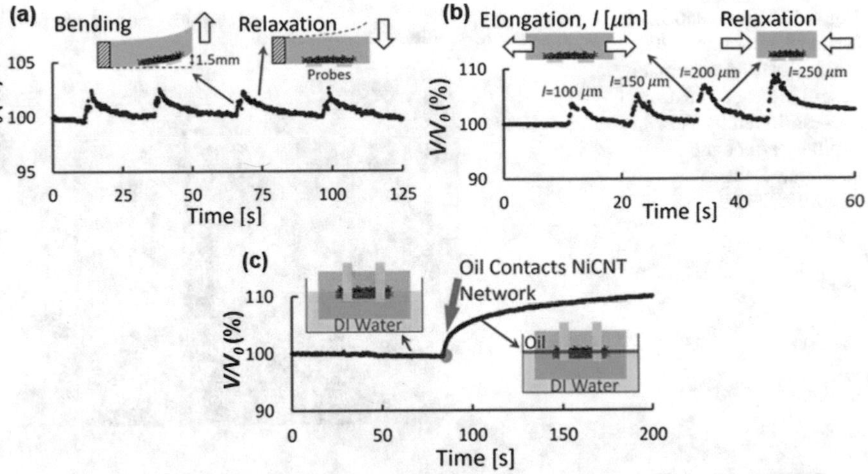

Fig. 2.11 The voltage drop V across the NiNP-CNT network rises when the block is subjected to (**a**) bending by displacement of the free end or (**b**) elongation of l = 100 μm, 150 μm, 200 μm, and 250 μm and (**c**) when the sensor is held at the air–water interface in a beaker and ~1 mL of oleic acid is dropped on the water, V rises when sensor contacts the oil. Reprinted (adapted) with permission from [110]. Copyright (2016) American Chemical Society

decreases. This increases the voltage, as shown in Fig. 2.11a, b. As the PDMS is relaxed, the distance between the interconnects reduces and the voltage gradually returns to its initial value. Continuous and cyclical mechanical deformation of the network leaves a permanent resistance change, as evident from Fig. 2.11b where the steady voltage on relaxation increases over the cycle. The printed NiNP-CNT network also responds to the presence of oil since the PDMS, being oleophilic, absorbs oil, affecting the percolation of the CNT networks and thus changing the resistance of the circuit (cf. Fig. 2.11c). This method of printing sensors with magnetic ink is made possible by combining the unique properties of MNPs and CNTs using non-covalent functionalization. The technique can be readily scaled up and integrated with existing nozzle-based printing of flexible circuitry and sensors that have complex geometries.

2.5 Endohedral Functionalization of CNT with MNPs

A CNT is a nano-sized container that has a protective carbon shell to encapsulate nanomaterials within. Thus a CNT is a smart nanoscale carrier that can be filled with tailored materials for target applications such as memory devices, optical transducers, wearable electronics, and medicine. MNPs can be intercalated inside CNTs during the synthesis process using the chemical vapor deposition of metallocenes [76, 115]. External molecules are encapsulated by a capillary effect inside

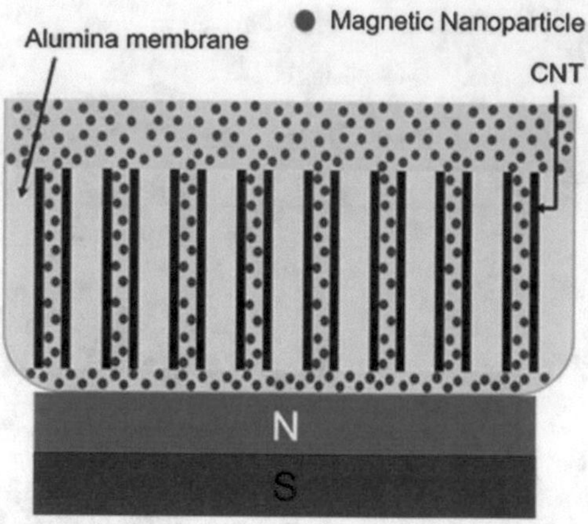

Fig. 2.12 Intercalation of magnetic nanomaterial inside CNTs. A schematic describing encapsulation process driven by the capillary effect and accelerated with a magnetic field

the CNTs [61]. MNPs can also be added to a CNT dispersion and encapsulated within the nanotubes. The solvent carrying MNPs wets the inside volume of CNTs and, after drying, the MNPs remain encapsulated as shown in Fig. 2.12.

2.6 Future Outlook

The covalent and noncovalent functionalization of CNTs with MNPs has produced a new class of nanoscale that exploits the unique properties of both CNTs and MNPs. MNP-functionalized CNTs can be used to prepare polymer composites and bulk materials with enhanced mechanical, electrical, and thermal properties. The properties can be tuned, depending on the direction of alignment of the CNT network. The materials can also potentially be used in energy storage applications such as supercapacitors [116] and lithium ion batteries. Functionalized CNTs have high sensitivity toward the respective external stimuli which could be used to develop custom cost-effective sensing platforms. Environmental sustainability demands the use of less hazardous functionalizing agents which could replace the acids that are currently used to functionalize CNTs. Crack patterns of the deposited magnetized CNTs need further attention which might reduce the critical percolation threshold for the sensing activity.

References

1. Iijima S (1991) Helical microtubules of graphitic carbon. Nature 354(6348):56–58
2. Iijima S, Ichihashi T (1993) Single-shell carbon nanotubes of 1-nm diameter. Nature 363(6430):603–605
3. Varshney K (2014) Carbon nanotubes: a review on synthesis, properties and applications. Int J Eng Res 2(4):660–677

4. Smalley RE, Dresselhaus MS, Dresselhaus G, Avouris P (2003) Carbon nanotubes: synthesis, structure, properties, and applications, vol 80. Springer Science & Business Media, Berlin
5. Saito R, Dresselhaus G, Dresselhaus MS (1998) Physical properties of carbon nanotubes, vol 35. World Scientific, Singapore
6. De Volder MF, Tawfick SH, Baughman RH, Hart AJ (2013) Carbon nanotubes: present and future commercial applications. Science 339(6119):535–539
7. Gou J, Lau K (2005) Modeling and simulation of carbon nanotube/polymer composites. In: A chapter in the handbook of theoretical and computational nanotechnology. American Scientific Publishers, Stevenson Ranch, pp 20051–58883
8. Mahar B, Laslau C, Yip R, Sun Y (2007) Development of carbon nanotube-based sensors—a review. IEEE Sensors J 7(2):266–284
9. Sinha N, Ma J, Yeow JT (2006) Carbon nanotube-based sensors. J Nanosci Nanotechnol 6(3):573–590
10. Popov VN, Lambin P (2006) Carbon nanotubes: from basic research to nanotechnology, vol 222. Springer Science & Business Media, Berlin
11. Peng L-M, Zhang Z, Wang S (2014) Carbon nanotube electronics: recent advances. Mater Today 17(9):433–442
12. Szabó A, Perri C, Csató A, Giordano G, Vuono D, Nagy JB (2010) Synthesis methods of carbon nanotubes and related materials. Materials 3(5):3092–3140
13. Antiohos D, Romano M, Chen J, Razal JM (2013) Carbon nanotubes for energy applications. In: Syntheses and applications of carbon nanotubes and their composites. InTech, Rijeka, pp 496–534
14. Wang G, Zhang L, Zhang J (2012) A review of electrode materials for electrochemical supercapacitors. Chem Soc Rev 41(2):797–828
15. Park S, Kim Y-S, Kim WB, Jon S (2009) Carbon nanosyringe array as a platform for intracellular delivery. Nano Lett 9(4):1325–1329
16. Shao W, Arghya P, Yiyong M, Rodes L, Prakash S (2013) Carbon nanotubes for use in medicine: potentials and limitations. In: Syntheses and applications of carbon nanotubes and their composites. InTech, Rijeka, pp 285–311
17. Sinha N, Yeow J-W (2005) Carbon nanotubes for biomedical applications. IEEE Trans Nanobiosci 4(2):180–195
18. Zhu Y, Xu C, Wang L (2011) Carbon nanotubes in biomedicine and biosensing. InTech Open Access, Rijeka
19. Prasek J, Drbohlavova J, Chomoucka J, Hubalek J, Jasek O, Adam V, Kizek R (2013) Chemical vapor depositions for carbon nanotubes synthesis. Nova Science, Hauppauge
20. Rafique MMA, Iqbal J (2011) Production of carbon nanotubes by different routes—a review. JEAS 1(02):29
21. Terranova ML, Sessa V, Rossi M (2006) The world of carbon nanotubes: an overview of CVD growth methodologies. Chem Vap Depos 12(6):315–325
22. Choudhary V, Gupta A (2011) Polymer/carbon nanotube nanocomposites. In: Carbon nanotubes-polymer nanocomposites, p 65–90
23. Du J, Bai J, Cheng H (2007) The present status and key problems of carbon nanotube based polymer composites. Express Polym Lett 1(5):253–273
24. Goh PS, Ismail AF, Ng BC (2014) Directional alignment of carbon nanotubes in polymer matrices: contemporary approaches and future advances. Compos A Appl Sci Manuf 56:103–126
25. Kimura T, Ago H, Tobita M, Ohshima S, Kyotani M, Yumura M (2002) Polymer composites of carbon nanotubes aligned by a magnetic field. Adv Mater 14(19):1380–1383
26. Homma Y, Chiashi S, Yamamoto T, Kono K, Matsumoto D, Shitaba J, Sato S (2013) Photoluminescence measurements and molecular dynamics simulations of water adsorption on the hydrophobic surface of a carbon nanotube in water vapor. Phys Rev Lett 110(15):157402

27. Lu SH, Ni Tun MH, Mei ZJ, Chia GH, Lim X, Sow C-H (2009) Improved hydrophobicity of carbon nanotube arrays with micropatterning. Langmuir 25(21):12806–12811
28. Gupta V, Kotnala RK (2012) Multifunctional ferromagnetic carbon-nanotube arrays prepared by pulse-injection chemical vapor deposition. Angew Chem Int Ed 51(12):2916–2919
29. Ganguly R, Gaind AP, Puri IK (2005) A strategy for the assembly of three-dimensional mesoscopic structures using a ferrofluid. Phys Fluids 17(5):057103
30. Ganguly R, Puri IK (2007) Field-assisted self-assembly of superparamagnetic nanoparticles for biomedical, MEMS and BioMEMS applications. Adv Appl Mech 41:293–335
31. Ganguly R, Puri IK (2010) Microfluidic transport in magnetic MEMS and bioMEMS. Wiley Interdiscip Rev Nanomed Nanobiotechnol 2(4):382–399
32. Puri IK, Ganguly R (2014) Particle transport in therapeutic magnetic fields. Annu Rev Fluid Mech 46:407–440
33. Roy T, Sinha A, Chakraborty S, Ganguly R, Puri IK (2009) Magnetic microsphere-based mixers for microdroplets. Phys Fluids 21(2):027101
34. Zhang B, Du Y, Zhang P, Zhao H, Kang L, Han X, Xu P (2013) Microwave absorption enhancement of Fe_3O_4/polyaniline core/shell hybrid microspheres with controlled shell thickness. J Appl Polym Sci 130(3):1909–1916
35. Teymourian H, Salimi A, Khezrian S (2013) Fe_3O_4 magnetic nanoparticles/reduced graphene oxide nanosheets as a novel electrochemical and bioeletrochemical sensing platform. Biosens Bioelectron 49:1–8
36. Ganguly R, Gaind AP, Puri IK (2004) Ferrofluid transport analysis for micro-and meso-scale applications. In: ASME 2004 International Mechanical Engineering Congress and Exposition. American Society of Mechanical Engineers, pp 65–68
37. Xu HL, Shen Y, Bi H (2012) Reduced graphene oxide decorated with Fe_3O_4 nanoparticles as high performance anode for lithium ion batteries. In: Key engineering materials. Trans Tech, pp 108–112
38. Frey NA, Peng S, Cheng K, Sun S (2009) Magnetic nanoparticles: synthesis, functionalization, and applications in bioimaging and magnetic energy storage. Chem Soc Rev 38(9):2532–2542
39. Wang G, Wang C, Dou W, Ma Q, Yuan P, Su X (2009) The synthesis of magnetic and fluorescent bi-functional silica composite nanoparticles via reverse microemulsion method. J Fluoresc 19(6):939–946
40. Han YC, Cha HG, Kim CW, Kim YH, Kang YS (2007) Synthesis of highly magnetized iron nanoparticles by a solventless thermal decomposition method. J Phys Chem C 111(17):6275–6280
41. Sato T, Iijima T, Seki M, Inagaki N (1987) Magnetic properties of ultrafine ferrite particles. J Magn Magn Mater 65(2):252–256
42. Chen D-H, He X-R (2001) Synthesis of nickel ferrite nanoparticles by sol-gel method. Mater Res Bull 36(7):1369–1377
43. Zhong Z, Lin M, Ng V, Ng GXB, Foo Y, Gedanken A (2006) A versatile wet-chemical method for synthesis of one-dimensional ferric and other transition metal oxides. Chem Mater 18(25):6031–6036
44. Polshettiwar V, Baruwati B, Varma RS (2009) Self-assembly of metal oxides into three-dimensional nanostructures: synthesis and application in catalysis. ACS Nano 3(3):728–736
45. Dosev D, Nichkova M, Dumas RK, Gee SJ, Hammock BD, Liu K, Kennedy IM (2007) Magnetic/luminescent core/shell particles synthesized by spray pyrolysis and their application in immunoassays with internal standard. Nanotechnology 18(5):055102
46. Ai L, Zhang C, Chen Z (2011) Removal of methylene blue from aqueous solution by a solvothermal-synthesized graphene/magnetite composite. J Hazard Mater 192(3):1515–1524
47. Du N, Xu Y, Zhang H, Zhai C, Yang D (2010) Selective synthesis of Fe_2O_3 and Fe_3O_4 nanowires via a single precursor: a general method for metal oxide nanowires. Nanoscale Res Lett 5(8):1295
48. Guo Q, Teng X, Rahman S, Yang H (2003) Patterned Langmuir-Blodgett films of monodisperse nanoparticles of iron oxide using soft lithography. J Am Chem Soc 125(3):630–631

49. Chia CH, Zakaria S, Yusoff M, Goh S, Haw C, Ahmadi S, Huang N, Lim H (2010) Size and crystallinity-dependent magnetic properties of $CoFe_2O_4$ nanocrystals. Ceram Int 36(2):605–609
50. Ciraci S, Dag S, Yildirim T, Gülseren O, Senger R (2004) Functionalized carbon nanotubes and device applications. J Phys Condens Matter 16(29):R901
51. Naseh MV, Khodadadi A, Mortazavi Y, Sahraei OA, Pourfayaz F, Sedghi SM (2009) Functionalization of carbon nanotubes using nitric acid oxidation and DBD plasma. World Acad Sci Eng Technol 49:177–179
52. Khare BN, Meyyappan M, Cassell AM, Nguyen CV, Han J (2002) Functionalization of carbon nanotubes using atomic hydrogen from a glow discharge. Nano Lett 2(1):73–77
53. Kooi SE, Schlecht U, Burghard M, Kern K (2002) Electrochemical modification of single carbon nanotubes. Angew Chem Int Ed 41(8):1353–1355
54. Lim JK, Yun WS, M-h Y, Lee SK, Kim CH, Kim K, Kim SK (2003) Selective thiolation of single-walled carbon nanotubes. Synth Met 139(2):521–527
55. Rinzler A, Liu J, Dai H, Nikolaev P, Huffman C, Rodriguez-Macias F, Boul P, Lu A, Heymann D, Colbert D (1998) Large-scale purification of single-wall carbon nanotubes: process, product, and characterization. Appl Phys A Mater Sci Process 67(1):29–37
56. Chen YK, Green ML, Griffin JL, Hammer J, Lago RM, Tsang SC (1996) Purification and opening of carbon nanotubes via bromination. Adv Mater 8(12):1012–1015
57. Riggs JE, Guo Z, Carroll DL, Sun Y-P (2000) Strong luminescence of solubilized carbon nanotubes. J Am Chem Soc 122(24):5879–5880
58. Huang W, Taylor S, Fu K, Lin Y, Zhang D, Hanks TW, Rao AM, Sun Y-P (2002) Attaching proteins to carbon nanotubes via diimide-activated amidation. Nano Lett 2(4):311–314
59. Georgakilas V, Kordatos K, Prato M, Guldi DM, Holzinger M, Hirsch A (2002) Organic functionalization of carbon nanotubes. J Am Chem Soc 124(5):760–761
60. Lin T, Bajpai V, Ji T, Dai L (2003) Chemistry of carbon nanotubes. Aust J Chem 56(7):635–651
61. Ma P-C, Siddiqui NA, Marom G, Kim J-K (2010) Dispersion and functionalization of carbon nanotubes for polymer-based nanocomposites: a review. Compos A Appl Sci Manuf 41(10):1345–1367
62. Balasubramanian K, Burghard M (2005) Chemically functionalized carbon nanotubes. Small 1(2):180–192
63. Chen L, Xie H, Yu W (2011) Functionalization methods of carbon nanotubes and its applications. InTech Open Access, Rijeka
64. Chen RJ, Bangsaruntip S, Drouvalakis KA, Kam NWS, Shim M, Li Y, Kim W, Utz PJ, Dai H (2003) Noncovalent functionalization of carbon nanotubes for highly specific electronic biosensors. Proc Natl Acad Sci U S A 100(9):4984–4989
65. Meng L, Fu C, Lu Q (2009) Advanced technology for functionalization of carbon nanotubes. Prog Nat Sci 19(7):801–810
66. Yang F, Jin C, Yang D, Jiang Y, Li J, Di Y, Hu J, Wang C, Ni Q, Fu D (2011) Magnetic functionalised carbon nanotubes as drug vehicles for cancer lymph node metastasis treatment. Eur J Cancer 47(12):1873–1882
67. Khan W, Sharma R, Saini P (2016) Carbon nanotube-based polymer composites: synthesis, properties and applications. Carbon nanotubes—current progress of their polymer composites. doi:50950
68. Hata K, Futaba DN, Mizuno K, Namai T, Yumura M, Iijima S (2004) Water-assisted highly efficient synthesis of impurity-free single-walled carbon nanotubes. Science 306(5700):1362–1364
69. Koziol K, Vilatela J, Moisala A, Motta M, Cunniff P, Sennett M, Windle A (2007) High-performance carbon nanotube fiber. Science 318(5858):1892–1895
70. Zhang M, Atkinson KR, Baughman RH (2004) Multifunctional carbon nanotube yarns by downsizing an ancient technology. Science 306(5700):1358–1361
71. Sathyanarayana S, Hübner C (2013) Thermoplastic nanocomposites with carbon nanotubes. In: Structural nanocomposites. Springer, Berlin, pp 19–60
72. Kim Y, Torrens ON, Kikkawa J, Abou-Hamad E, Goze-Bac C, Luzzi DE (2007) High-purity diamagnetic single-wall carbon nanotube buckypaper. Chem Mater 19(12):2982–2986

73. Ovchinnikov AA (1994) Giant diamagnetism of carbon nanotubes. Phys Lett A 195(1):95–96
74. Das A, Raffi M, Megaridis C, Fragouli D, Innocenti C, Athanassiou A (2015) Magnetite (Fe_3O_4)-filled carbon nanofibers as electro-conducting/superparamagnetic nanohybrids and their multifunctional polymer composites. J Nanopart Res 17(1):1–14
75. Korneva G, Ye H, Gogotsi Y, Halverson D, Friedman G, Bradley J-C, Kornev KG (2005) Carbon nanotubes loaded with magnetic particles. Nano Lett 5(5):879–884
76. Sameera I, Bhatia R, Prasad V (2013) Characterization and magnetic response of multiwall carbon nanotubes filled with iron nanoparticles of different aspect ratios. Physica E 52:1–7
77. Weissker U, Hampel S, Leonhardt A, Büchner B (2010) Carbon nanotubes filled with ferromagnetic materials. Materials 3(8):4387–4427
78. He H, Gao C (2011) Synthesis of Fe_3O_4/Pt nanoparticles decorated carbon nanotubes and their use as magnetically recyclable catalysts. J Nanomater 2011:11
79. Zhang Q, Zhu M, Zhang Q, Li Y, Wang H (2009) Synthesis and characterization of carbon nanotubes decorated with manganese–zinc ferrite nanospheres. Mater Chem Phys 116(2):658–662
80. Zhao W, Zhu L, Lu Y, Zhang L, Schuster RH, Wang W (2013) Magnetic nanoparticles decorated multi-walled carbon nanotubes by bio-inspired poly (dopamine) surface functionalization. Synth Met 169:59–63
81. Xiao D, Dramou P, Xiong N, He H, Li H, Yuan D, Dai H (2013) Development of novel molecularly imprinted magnetic solid-phase extraction materials based on magnetic carbon nanotubes and their application for the determination of gatifloxacin in serum samples coupled with high performance liquid chromatography. J Chromatogr A 1274:44–53
82. Huang Y, Yuan Y, Zhou Z, Liang J, Chen Z, Li G (2014) Optimization and evaluation of chelerythrine nanoparticles composed of magnetic multiwalled carbon nanotubes by response surface methodology. Appl Surf Sci 292:378–386
83. Yu P, Ma H, Shang Y, Wu J, Shen S (2014) Polyethylene glycol modified magnetic carbon nanotubes as nanosorbents for the determination of methylprednisolone in rat plasma by high performance liquid chromatography. J Chromatogr A 1348:27–33
84. Chang PR, Zheng P, Liu B, Anderson DP, Yu J, Ma X (2011) Characterization of magnetic soluble starch-functionalized carbon nanotubes and its application for the adsorption of the dyes. J Hazard Mater 186(2):2144–2150
85. Fazelirad H, Ranjbar M, Taher MA, Sargazi G (2015) Preparation of magnetic multi-walled carbon nanotubes for an efficient adsorption and spectrophotometric determination of amoxicillin. J Ind Eng Chem 21:889–892
86. Sowichai K, Supothina S, Nimittrakoolchai O-u, Seto T, Otani Y, Charinpanitkul T (2012) Facile method to prepare magnetic multi-walled carbon nanotubes by in situ co-precipitation route. J Ind Eng Chem 18(5):1568–1571
87. Mitróová Z, Tomašovičová N, Lancz G, Kováč J, Vávra I, Kopčanský P (2010) Preparation and characterization of carbon nanotubes functionalized by magnetite nanoparticles. Olomouc, Czech Republic, EU 10, pp 12–14
88. Zarei H, Ghourchian H, Eskandari K, Zeinali M (2012) Magnetic nanocomposite of anti-human IgG/COOH–multiwalled carbon nanotubes/Fe_3O_4 as a platform for electrochemical immunoassay. Anal Biochem 421(2):446–453
89. Abdalla AM, Ghosh S, Puri IK (2016) Decorating carbon nanotubes with co-precipitated magnetite nanocrystals. Diam Relat Mater 66:90–97
90. Ghosh S, Puri IK (2015) Changing the magnetic properties of microstructure by directing the self-assembly of superparamagnetic nanoparticles. Faraday Discuss 181:423–435
91. Kappiyoor R, Liangruksa M, Ganguly R, Puri IK (2010) The effects of magnetic nanoparticle properties on magnetic fluid hyperthermia. J Appl Phys 108(9):094702
92. Sarojini KK, Manoj SV, Singh PK, Pradeep T, Das SK (2013) Electrical conductivity of ceramic and metallic nanofluids. Colloids Surf A Physicochem Eng Asp 417:39–46
93. Karthik P, Singh SP (2015) Conductive silver inks and their applications in printed and flexible electronics. RSC Adv 5(95):77760–77790
94. Glover B, Whites KW, Hong H, Mukherjee A, Billups WE (2008) Effective electrical conductivity of functional single-wall carbon nanotubes in aqueous fluids. Synth Met 158(12):506–508

95. Eapen J, Rusconi R, Piazza R, Yip S (2010) The classical nature of thermal conduction in nanofluids. J Heat Transf 132(10):102402
96. Kamyshny A, Magdassi S (2014) Conductive nanomaterials for printed electronics. Small 10(17):3515–3535
97. Kraynov A, Müller TE (2011) Concepts for the stabilization of metal nanoparticles in ionic liquids. InTech Open Access, Rijeka
98. Salazar PF, Chan KJ, Stephens ST, Cola BA (2014) Enhanced electrical conductivity of imidazolium-based ionic liquids mixed with carbon nanotubes: a spectroscopic study. J Electrochem Soc 161(9):H481–H486
99. Othman SH, Rashid SA, Ghazi TIM, Abdullah N (2012) Dispersion and stabilization of photocatalytic TiO_2 nanoparticles in aqueous suspension for coatings applications. J Nanomater 2012:2
100. Moore VC, Strano MS, Haroz EH, Hauge RH, Smalley RE, Schmidt J, Talmon Y (2003) Individually suspended single-walled carbon nanotubes in various surfactants. Nano Lett 3(10):1379–1382
101. Voit W, Kim D, Zapka W, Muhammed M, Rao K (2001) Magnetic behavior of coated superparamagnetic iron oxide nanoparticles in ferrofluids. In: MRS proceedings. Cambridge University Press, p Y7.8
102. Abdalla AM, Fattah ARA, Ghosh S, Puri IK (2017) Magnetoresponsive conductive colloidal suspensions with magnetized carbon nanotubes. J Magn Magn Mater 421:292–299
103. Mozaffari M, Amighian J, Tavakoli R (2015) The effect of yttrium substitution on the magnetic properties of magnetite nanoparticles. J Magn Magn Mater 379:208–212
104. Murakami K (1965) The characteristics of ferrite cores with low Curie temperature and their application. IEEE Trans Magn 1(2):96–100
105. Rezlescu N, Cuciureanu E (1970) Cation distribution and Curie temperature in some ferrites containing copper and manganese. Phys Status Solidi A 3(4):873–878
106. Mahanthesha P, Srinivasa C, Mohankumar G (2014) Processing and characterization of carbon nanotubes decorated with pure electroless nickel and their magnetic properties. Proc Mater Sci 5:883–890
107. Chen X, Xia J, Peng J, Li W, Xie S (2000) Carbon nanotube metal matrix composites pre pared by electroless plating. Compos Sci Technol 60(2):301–306
108. Abdalla AM, Majdi T, Ghosh S, Puri IK (2016) Fabrication of nanoscale to macroscale nickel-multiwall carbon nanotube hybrid materials with tunable material properties. Mater Res Express 3(12):125014
109. Bittencourt C, Felten A, Ghijsen J, Pireaux J, Drube W, Erni R, Van Tendeloo G (2007) Decorating carbon nanotubes with nickel nanoparticles. Chem Phys Lett 436(4):368–372
110. Abdel Fattah AR, Majdi T, Abdalla AM, Ghosh S, Puri IK (2016) Nickel nanoparticles entangled in carbon nanotubes: novel ink for nanotube printing. ACS Appl Mater Interfaces 8(3):1589–1593
111. Zhang B, Dong Q, Korman CE, Li Z, Zaghloul ME (2013) Flexible packaging of solid-state integrated circuit chips with elastomeric microfluidics. Sci Rep 3
112. Amjadi M, Yoon YJ, Park I (2015) Ultra-stretchable and skin-mountable strain sensors using carbon nanotubes? Ecoflex nanocomposites. Nanotechnology 26(37):375501
113. Lu J, Zhang X, Wu X, Dai Z, Zhang J (2015) A Ni-doped carbon nanotube sensor for detecting oil-dissolved gases in transformers. Sensors 15(6):13522–13532
114. Wang J (2005) Carbon-nanotube based electrochemical biosensors: a review. Electroanalysis 17(1):7–14
115. Müller C, Golberg D, Leonhardt A, Hampel S, Büchner B (2006) Growth studies, TEM and XRD investigations of iron-filled carbon nanotubes. Phys Status Solidi A 203(6):1064–1068
116. Abdalla A, Sahu R, Wallar C, Chen R, Zhitomirsky I, Puri I (2016) Nickel oxide nanotubes synthesis using multiwalled carbon nanotubes as sacrificial templates for supercapacitor application. Nanotechnology 28(7):075603. doi:10.1088/1361-6528/aa53f3

Part II
Two Dimensional Nanomaterials

Chapter 3
2D Materials: Applications for Electrochemical Energy Storage Devices

Shan Hu, Suprem R. Das, and Hosein Monshat

3.1 Introduction

3.1.1 Brief Introduction of Electrochemical Energy Storage

As sustainable energy devices and systems are being emphasized and projected for addressing one of the most important grand challenges of the twenty-first century, research efforts on the two energy storage devices such as supercapacitor and battery, in conjunction with tremendous material discovery and innovation, have become ever intense and multidisciplinary. The motivation is largely to unify both the energy storage devices and to build higher energy efficient device and system for sustainable, clean energy applications. Fundamentally, while in a (electrolytic) capacitor the energy is stored electrostatically, a battery stores it via reversible electrochemical

S. Hu (✉)
Mechanical Engineering Department, Iowa State University,
2025 Black Engineering, Ames, IA 50011, USA

Microelectronics Research Center, Iowa State University,
Applied Sciences Complex, Ames, IA 50011, USA
e-mail: shanhu@iastate.edu

S.R. Das
Mechanical Engineering Department, Iowa State University,
2025 Black Engineering, Ames, IA 50011, USA

Microelectronics Research Center, Iowa State University,
Applied Sciences Complex, Ames, IA 50011, USA

Division of Materials Science and Engineering, Ames Laboratory, Ames, IA 50011, USA

H. Monshat
Mechanical Engineering Department, Iowa State University,
2025 Black Engineering, Ames, IA 50011, USA

G. Balasubramanian (ed.), *Advances in Nanomaterials*,
DOI 10.1007/978-3-319-64717-3_3

reactions. Consequently, while a battery charges and discharges slowly (low power), it can store more energy per unit mass or unit volume of active materials (high energy density). On the other hand, a conventional (electrolytic) capacitor has high power but low energy density. Figure 3.1 shows the comparison between a capacitor and battery from these two metrics perspective (called a "Ragone Plot") [1]. So ideally, an energy storage device with high power density (powers per volume, or powers per mass) as well as high energy density (energy per volume, or energies per mass) would unify the above two storage devices. Such an attempt along with multifunctional materials and state-of-the-art device architecture would transform the new generation of devices toward high-energy and high-power applications.

Besides batteries and electrolytic capacitors, an emerging energy storage mechanism, i.e., supercapacitors have demonstrated the potential for high-power and high-energy storage. Compared with batteries, supercapacitors can provide higher levels of electrical power and offer longer operating lifetimes. Furthermore, supercapacitors experience no memory effect and are safer compared with batteries. They can deliver much more energy density than electrolytic capacitors. Electrostatic double-layer and pseudocapacitance are two main distinguished mechanisms for charge storage in supercapacitors. Pseudocapacitors that could undergo highly reversible surface redox reactions are able to store more charges than regular double-layer capacitors. High surface redox reactions rate brings both high energy density of lithium ion batteries as well as high power density of capacitors for supercapacitors. 2D nanomaterials have recently been studied much due to their unique characteristics and advantages as supercapacitor electrode materials.

3.1.2 Fundamentals of Energy Storage in Supercapacitors

Like a battery, supercapacitor consists of three essential components such as electrodes, electrolyte, and separator. Based on unique charge storage mechanisms, the supercapacitors (SCs) can be divided into three categories: electrochemical double layer capacitors (EDLCs), pseudocapacitors (PCs), and hybrid capacitors (HCs). While an EDLC uses a non-Faradic mechanism (no chemical reaction, it involves purely physical processes such as surface physisorption), a PC uses a Faradic process (transfer of charge along the interface between electrolyte and electrode, associated with an oxidation–reduction reaction called a reversible redox reaction) and a HC uses both the mechanisms during the device operation.

3.1.2.1 Electric Double Layer Capacitors

Electric double layer capacitors (EDLC) rely on surface physisorption for charge storage. Figure 3.2 is the schematic of an EDLC during charging: anions and cations in the electrolyte accumulate at the electrode/electrolyte interfaces forming two electric double layers. Each electric double layer is equivalent to a capacitor, with capacitance given in Eq. (3.1)

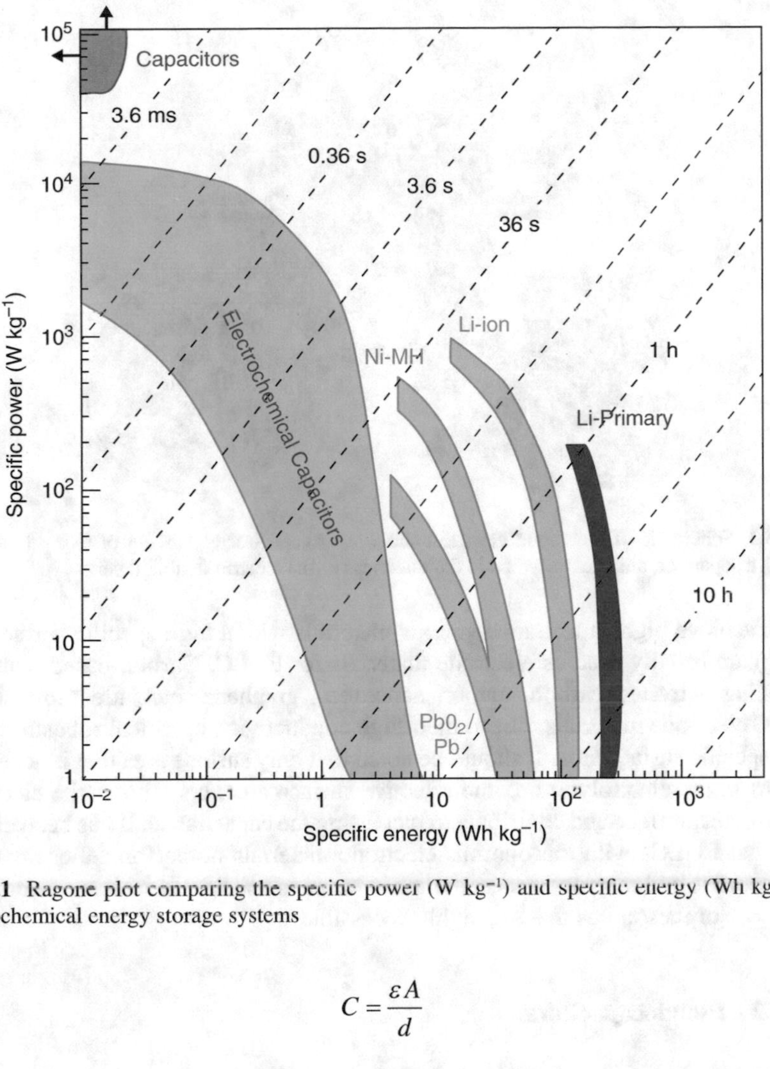

Fig. 3.1 Ragone plot comparing the specific power (W kg^{-1}) and specific energy (Wh kg^{-1}) of electrochemical energy storage systems

$$C = \frac{\varepsilon A}{d} \tag{3.1}$$

where ε is the dielectric constant of the electrolyte, A is the surface area of the electrode/electrolyte interface, and d is the thickness of the electric double layer. The thickness depends on the concentration of the electrolyte and the size of electrolyte ions. With concentrated electrolyte, the thickness value is usually in the range of 0.2–1 nm. Total capacitance of a supercapacitor cell (C_{Cell}) consisting of a positive and a negative electrode is given by Eq. (3.2)

$$\frac{1}{C_{Cell}} = \frac{1}{C_{PE}} + \frac{1}{C_{NE}} \tag{3.2}$$

where C_{PE} and C_{NE} are the capacitances of the positive electrode and negative electrode respectively.

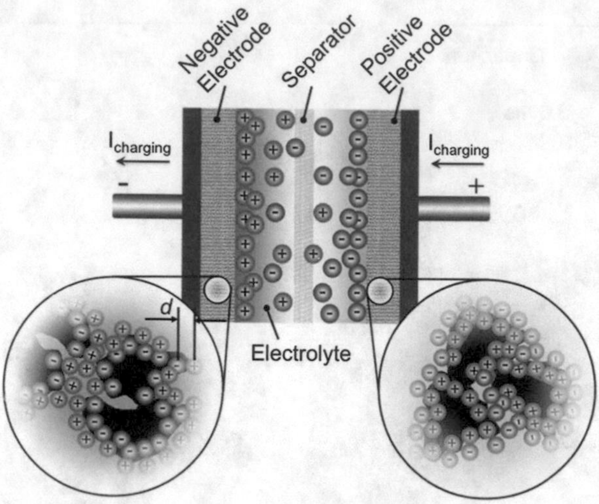

Fig. 3.2 Schematic of a charged electric double layer capacitor consisting of two porous electrodes, a separator, and electrolyte: *d* is the thickness of the electric double layer

To achieve high capacitance, porous materials with a high specific surface area (m^2/g) are usually used as electrode materials for EDLC. Carbon-based materials, including activated carbon, carbon nanotubes, graphene, etc., are most studied EDLC electrode materials, due to their high conductivity, chemical robustness, and high specific surface area. It should be noted that only surface area that is accessible to ions in the electrolyte, i.e., the effective surface area, constitutes the electrode/electrolyte interface and contribute to increasing the capacitance. It has been discovered that EDLCs with mesoporous electrode materials outperform those made of microporous and macroporous materials in terms of specific capacitance, due to the existence of mesopores that are highly accessible.

3.1.2.2 Pseudocapacitors

In pseudocapacitors, surface physisorption and redox reaction work simultaneously for charge storage.

Figure 3.3 shows the charge storage at the anode and cathode of a pseudocapacitor during charging. Similar to an EDLC, electric double layers build up at the electrode/electrolyte interface. What is unique to pseudocapacitor is that charges can cross the electrode/electrolyte interface to participate in redox reaction at the electrodes. Due to the coexistence of two storage mechanisms, the total capacitance of a pseudocapacitor includes faradic capacitance (capacitance due to redox reaction) and the electric double layer capacitance. The former is usually more than an order of magnitude higher than the later. Electrode materials for pseudocapacitors include conducting polymers and transition metal oxides, for example, RuO_2, MnO_2, Co_3O_4, etc. The redox reaction storage mechanism of pseudocapacitors is similar to that of batteries. The difference is that in pseudocapacitor redox reactions

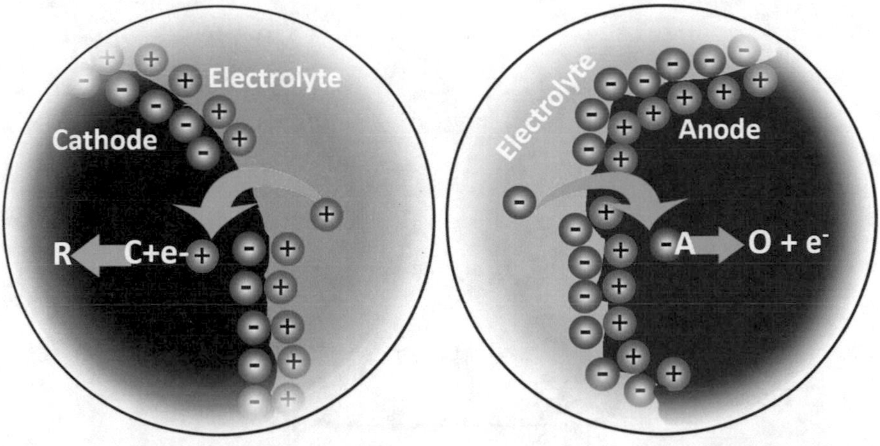

Fig. 3.3 Schematic of the anode and cathode of a pseudocapacitor during charging, showing the electric double layer at the electrode/electrolyte interface and the charge transport across this interface (A: anode material. C: cathode material. R: product of reduction reaction. O: product of oxidation reaction. e^-: electron)

take place at the surface or near the surface of the electrode, whereas in batteries reactions in bulk electrode dominate. Since surface/near-surface reaction is faster than bulk reaction, pseudocapacitors have higher charging/discharging speed than batteries.

3.1.2.3 Asymmetric Hybrid Capacitors

EDLC and pseudocapacitor each have their own pros and cons. Pseudocapacitors have higher energy density, whereas EDLCs have higher power rate (i.e., charging/discharging rate) and better cyclic stability. To combine the advantages of EDLCs and pseudocapacitors, asymmetrical hybrid capacitors have been proposed. In an asymmetrical cell, one electrode relies on EDL for charge storage and the other electrode is made of electrode materials that undergo redox reaction with the electrolyte or electrolyte that undergoes redox reaction at the surface of the electrode (Fig. 3.4).

3.2 2D Materials for Electrochemical Energy Storage

Graphene and molybdenum disulfide (MoS_2), two classic 2D materials with unprecedented thickness control down to single atomic layer and/or unit cell layer, have shown a long list of rich physical, electrical, mechanical, optical, thermal, and chemical properties. These, in turn, have shown the tremendous possibilities of their usage in a wide range of applications. Although both of them have 2D molecular structures, fundamentally they have the following differences: (1) graphene consists of a one atom thin, honeycomb lattice of sp^2 bonded carbon atoms (it is an organic

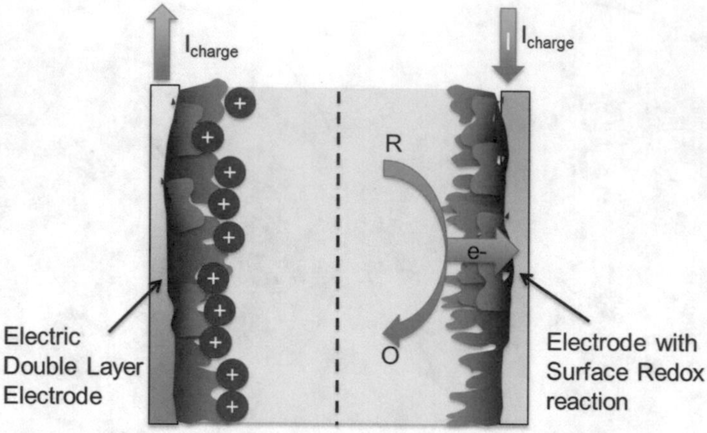

Fig. 3.4 Schematic of the anode and cathode of a hybrid supercapacitor during charging, showing the electric double layer forming at the cathode and the surface redox reaction at the anode (R: product of reduction reaction. O: product of oxidation reaction. e^-: electron)

membrane) whereas MoS_2 is often expressed as *inorganic-analog of graphene* that consists of single atomic sheet of molybdenum sandwiched between two atomic layers of sulfur (henceforth referred as single unit-cell layer); (2) the associated electronic band structure—whereas graphene has *linear* band structure with exceptionally high electron Fermi velocity ($V_F \sim c/300$, with c being speed of light in vacuum) and mobility, MoS_2 is more of a semiconductor with *parabolic* band structure and with a finite bandgap (1.2 eV bulk indirect bandgap that transitions to 1.8 eV of direct bandgap in monolayer limit). Lack of electronic bandgap in graphene, although provides low effective mass and high carrier mobility, makes it unsuccessful for demonstrating efficient electronic switch whereas MoS_2 becomes a candidate of choice for the future electronic switch. Figure 3.5 shows the schematic diagram of single layer/single unit-cell layer of each of these two 2D crystals with associated band structures (band structure of MoS_2 is shown in 2H crystal symmetry, 2H-MoS_2 is semiconducting). However, the common feature between these two materials is their ability to be isolated down to single atomic layers starting from bulk form due to weakly held van der Waal's interlayer coupling. Recently, 1T-MoS_2 with excellent electrical conductivity has been discovered as a separate phase material with the potential of supercapacitor applications [2].

Although it is not fully clear at this point for a possible pathway toward commercialization and the field is still in its infancy, the energy storage performance of graphene, MoS_2 and other 2D materials in transition metal dichalcogenide (TMD) family provide a rich testbed for a potential replacement of existing technologies due to the following reasons: (1) they possess layered van der Waal stacked structures with promising electronic and mechanical properties which allow electroactive species (ions) such as lithium, sodium, potassium, iodine, etc., to intercalate and deintercalate from the structure; (2) tremendous flexibilities in combining them with

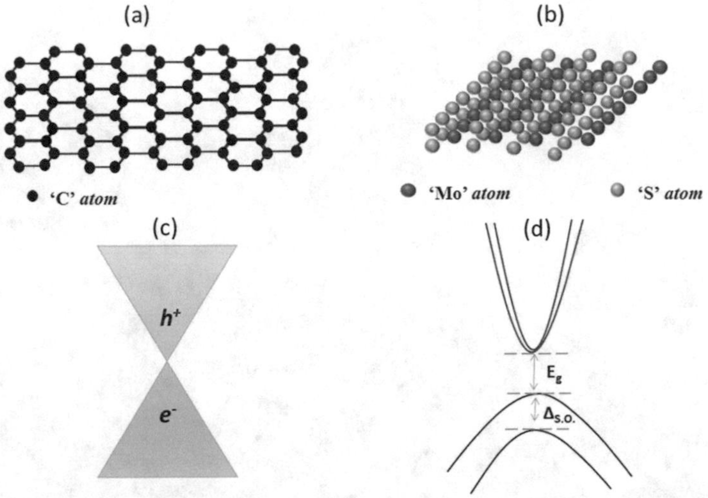

Fig. 3.5 The schematic diagram of the monolayers of (**a**) graphene and (**b**) molybdenum disulfide. In a typical graphite (HOPG) and MoS_2 crystal, these atomic layers are stacked vertically by van der Waal interaction, making these suitable for atomic layer isolation; (**c**) The "linear" band structure of single layer graphene (SLG) with sharp cone at the center called the "Dirac point". It is because of this special band structure of graphene, it acquires numerous special characteristics in electrical, optical, mechanical, chemical, and thermal properties; (**d**) The parabolic, direct band structure of 2H-MoS_2 single layer with electronic band gap along K-Γ crystal direction

other nanomaterials and nanostructures for exploring composite/hybrid structures with multifunctional properties; (3) exploiting bottom-up approaches to achieve various nano-architecture for enhanced functionalities; (4) easy functionalization of various foreign adsorbates to combined manipulation of surface enhanced properties and performance reliability; and (5) high theoretical specific surface area of 2000 m^2/g.

3.2.1 2D Materials for Supercapacitor Applications

Graphene has been used by numerous researchers as electrode materials for supercapacitor, mainly due to its excellent electrical conductivity, high surface area, and ability to undergo surface functionalization. Applications can be found in electric double layer capacitors, pseudocapacitors, and hybrid capacitors.

As discussed in Eq. (3.1), the capacitance of an EDLC depends on the specific surface area (SSA) of electrode materials and the effective SSA is the surface accessible to ions in the electrolyte. With graphene's theoretical SSA of 2000 m^2/g, the ideal attainable capacitance should be 500 F/g. However, the practically obtained values are only half of this ideal value at the best. The reason is that part of the surface area of graphene is from micropores (with pore diameter < 2 nm), which is often hard or non-accessible for ions. To increase the effective SSA, researchers at

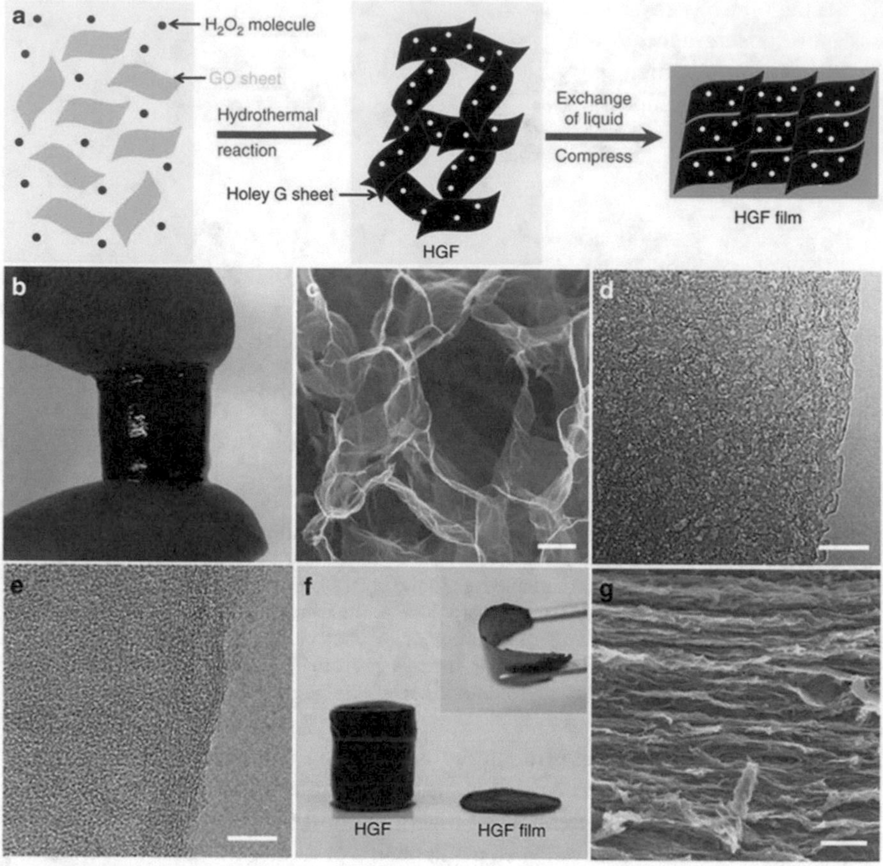

Fig. 3.6 Holey graphene framework (HGF) for high-performance supercapacitors. (**a**) Schematic of fabrication process of the HGF and HGF films. (**b**) a photograph of the as-fabricated HGF. (**c**) SEM image of the porous structure of the HGFs. Scale bar, 1 μm. (**d**, **e**) TEM image of the HGF (**d**) and non-holey graphene (**e**), showing many more mesopores in the HGF than non-holey counterpart. Scale bar, 10 nm. f. A photograph showing HGFs before and after mechanical compression with the flexibility of the compressed HGF film shown in the inset. (**g**) Cross-sectional SEM image of the compressed HGF film. Scale bar, 1 μm

University of California Los Angeles [3] have developed a method to create mesopores in graphene by etching the graphene oxide (a precursor for graphene) with hydrogen peroxide to create mesopores and later reduced the "holey" graphene oxide into holey graphene framework (HGF) (Fig. 3.6). The HGF was later compressed into HGF film and directly used as supercapacitor electrodes. The specific capacitance 278 F/g achieved from HGF-based supercapacitor was among the best reported so far for graphene based EDLC, demonstrating the importance of mesoporosity for EDL-based storage. A fully packaged device stack can deliver gravimetric and volumetric energy densities of 35 Wh kg^{-1} and 49 Wh L^{-1}, respectively, approaching those of lead acid batteries.

Besides activating the surface of graphene to increase effective SSA, researchers also developed graphene composite with other carbon nanomaterials such as activated carbon and carbon nanotube. Zheng et al. [4] synthesized graphene/activated carbon nanosheet composite with high SSA (2106 $m^2 g^{-1}$) for making supercapacitor electrode, exhibiting high specific capacitance up to 210 F g^{-1} in aqueous electrolyte. Its energy density can reach 22.3 Wh kg^{-1}, which is much higher than that of conventional supercapacitors based on activated carbon (5–6 Wh kg^{-1}).

Another strategy to enhance the specific capacitance of graphene-based supercapacitor besides pursuing large surface area is adding redox active species to the graphene electrode to introduce pseudocapacitance for charge storage. Pseudocapacitance can be enabled by surface functional groups of graphene. Common redox reaction induced by surface functional groups are summarized in Table 3.1. It should be noted that these functional groups only have redox activity under the appropriate electrolyte pH value. Among them, carboxylic groups and hydroxyl groups undergo redox reaction as shown in Table 3.1 in basic electrolyte (pH > 7). Pyrone-like groups, Quinone-like groups, and Chromene-like groups show redox activity in acidic electrolyte (pH < 7). Lactone group is a special case which undergoes redox reaction in both acidic and basic electrolytes.

Besides surface functionalization, pseudocapacitance can be enabled in graphene-based supercapacitor by loading the graphene with redox-active nanomaterials, including transition metal oxide, hydroxides, (oxy)hydroxide, and conductive polymer. In the resulting nanocomposite, graphene acts as conductive framework that provides direct electron transport path to/from the redox-active additives.

Wang et al. directly synthesized $Ni(OH)_2$ nanoplates onto graphene via a hydrothermal method and the resulting powder-form nanocomposite was made into electrode using polytetrafluoroethylene as binder. The nanocomposite electrode was tested in a three-electrode system with the $Ni(OH)_2$/graphene as working electrode, Ag/AgCl as reference electrode, and platinum wire as counter electrode. High specific capacitance of ~1335 F g^{-1} was obtained, which is much higher than the specific capacitance of a graphene-based EDLC [10]. However, it should be noted that the 1335 F g^{-1} represents the electrode's specific capacitance, rather than the cell-level capacitance. For practical application, what matters is the cell-level capacitance, i.e., the C_{Cell}. As given in Eq. (3.2), cell-level capacitance depends on the capacitance of both positive electrode and negative electrode. To achieve high C_{Cell}, the graphene/ $Ni(OH)_2$ needs a matching positive or negative electrode with equally high or even higher specific capacitance. Similar graphene nanocomposites for pseudocapacitor have been reported for V_2O_5, RuO_2, MnO_2, Co_2O_3/Co_3Co_4, etc. [11–15]. In all cases, electrode-level-specific capacitances are consistently higher than that of EDLC.

Compared with graphene, MoS_2 is used less often as active materials for supercapacitors mainly due to the fact that MoS_2 is a semiconducting material with poor electrical conductivity. Nevertheless, when integrated with conductive materials, MoS_2 still provides several advantages as active materials for supercapacitor electrodes: (1) its 2D structure provides the large surface areas; (2) the molybdenum centers of MoS_2 allow for strong coordination with nitrogen atoms in conducting polymers such as polyaniline (PANI) or polypyrrole (PPy), which benefits control-

Table 3.1 Surface functional groups with redox activity [5–9]

Name	Redox reaction mechanism
Surface groups with redox activity in basic electrolyte	
Carboxylic groups	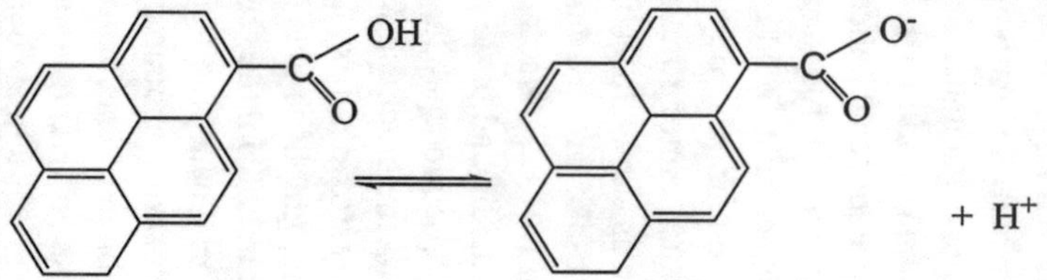
Hydroxyl groups	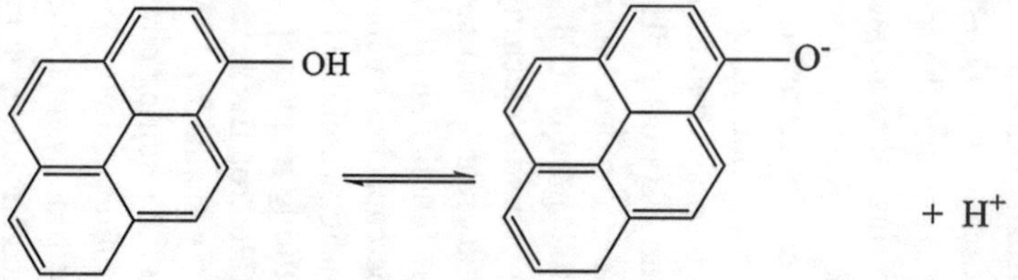

Name	Redox reaction mechanism
Surface groups with redox activity in acidic electrolyte	
Pyrone-like groups	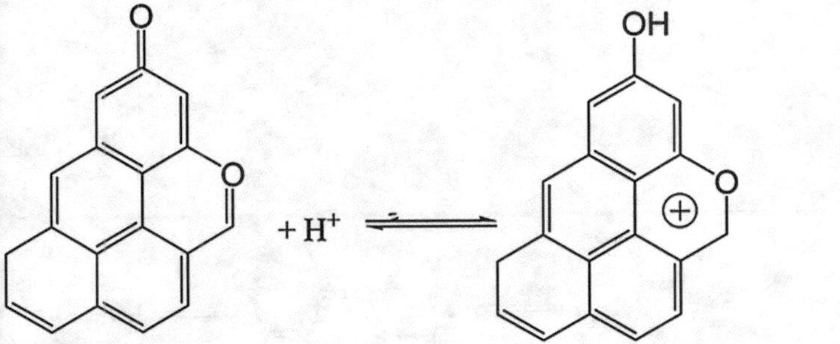
Quinone-like groups	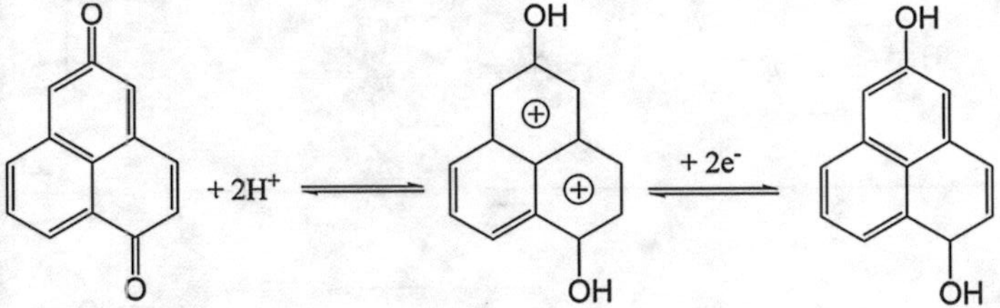
Chromene-like groups	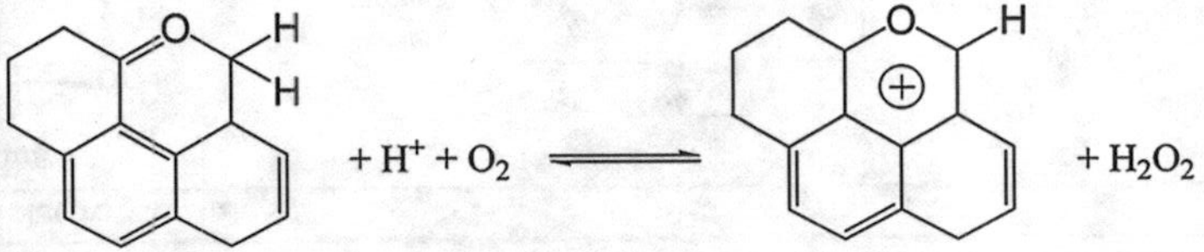

Table 3.1 (continued)

Name	Redox reaction mechanism
Surface groups with redox activities in both acidic and basic electrolyte	
Lactones	In acidic medium 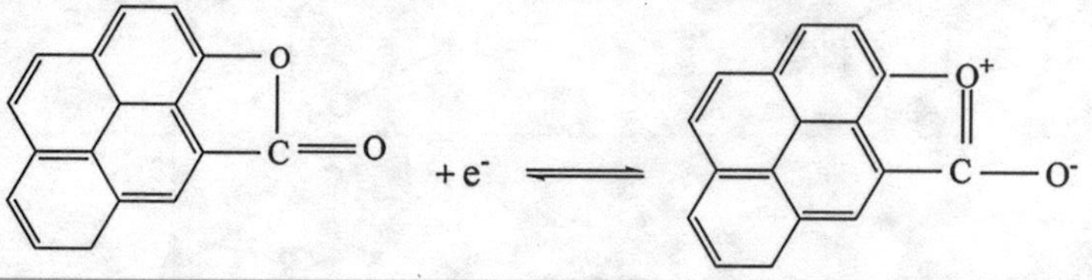
	In basic medium 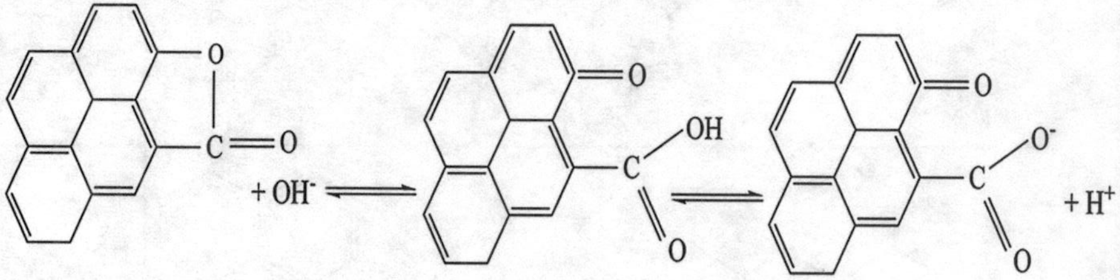

lable growth of conducting polymers onto the 2D nanosheet surfaces; (3) the Mo element in MoS_2 nanosheets possesses a range of oxidation states from +2 to +6, and could undergo reversible redox reactions, which give rise to additional pseudocapacitance. Therefore, a combination of MoS_2 with conductive polymers is a rational choice for developing supercapacitor electrodes with exceptional energy density and power density. Tang et al. [16] started with natural MoS_2 crystals which consist of stacks of single-layer MoS_2 nanosheets and used subsequent lithium intercalation and ultrasonication to exfoliate with natural MoS_2 into single layer MoS_2 nanosheets to fully expose the surface area of MoS_2. Then the exfoliated MoS_2 nanosheets were added into pyrrole monomers and cooled down to 5 °C while stirring for 30 min, followed by polymerization for 12 h. The product was collected by filtration and mixed with carbon black and PTFE binders to produce supercapacitor electrodes. A symmetric supercapacitor with exactly the same MoS_2/PPy electrodes was built and a specific capacitance of 695 F g^{-1} was achieved at the cell level (capacitance normalized to the total mass of active materials at two electrodes). With a cell voltage window of 0.9 V, energy density of 83.3 Wh kg^{-1} was achieved at a power density of 3332 W kg^{-1}. Interestingly, this work also prepared graphene/PPy nanocomposite and demonstrated that the electrochemical performance of MoS_2/PPy electrode is better than that of graphene/PPy, possibly due to the fact that the graphene was not exfoliated and hence stacking of graphene could lead to the loss of effective specific surface area in the graphene/PPy nanocomposite.

3.3 Challenges and Outlook

There have been considerable efforts in the last decade on 2D materials (graphene and beyond graphene) research. It includes fundamental research understanding the nanoscale physics of these materials to translating them for enormous prototype laboratory scale applications. When it comes to the research reports, technology disclosures, as well as start-up companies, 2D graphene, MoS_2, and other members of TMD are top recognized for their energy storage and energy conversion applications. In spite of this incredibly large efforts and progresses on energy storage applications for graphene, MoS_2, and other members of TMD family, reports on their pathway toward large-scale industrial prototyping and commercialization are sparse and a benchmarking with existing technology is rare. In our belief, in addition to the roadmap of 2D materials, an independent roadmap and a benchmarking for 2D material energy storage device are of prime importance. Due to sp^2 bonding in planar hexagonal honeycomb carbon, exceptionally good electronic conductivity (for transport) and exceptionally good thermal conductivity (for thermal managements), graphene is likely going to continue as the major candidate for innovation and graphene-based supercapacitors have more likelihood to reach commercialization stage in near future.

3.3.1 Challenges

Ongoing research on 2D materials continues to demonstrate improved energy storage performances. However, most of these 2D material-based systems still only exist in research labs. Several issues need to be resolved by researchers in this field to promote the wider adoption of 2D materials into practical applications. The first issue is the cost of material. The price of high-purity graphene and MoS_2 is still far from cost-effective, mainly due to the lack of effective approach to synthesize them in scalable quantities. The second issue is the lacking of understanding on the environmental and health effect of 2D materials and the absence of regulation. Recently, much research effort has been devoted to close this knowledge gap and the regulation regarding the use of 2D materials is expected to be established based on the research findings. The third issue is the stability of 2D materials, particularly for MoS_2 and other members of TMD. Although they have the capability of hosting other ions to intercalate, it is at the expense of their stability in the crystal structure. Repeated charging and discharging of the devices employing solely these materials leads to structural instability and poor performance (e.g., poor power density and cycling in supercapacitors, poor energy density, and cycling in battery). Effective approaches to stabilize the 2D materials are an ongoing research effort.

References

1. Simon P, Gogotsi Y (2008) Materials for electrochemical capacitors. *Nat. Mater.* 7(11):845–854
2. Acerce M, Voiry D, Chhowalla M (2015) Metallic 1T phase MoS_2 nanosheets as supercapacitor electrode materials. Nat Nanotechnol 10(4):313–318
3. Xu Y, Lin Z, Zhong X, Huang X, Weiss NO, Huang Y, Duan X (2014) Holey graphene frameworks for highly efficient capacitive energy storage. Nat Commun 5:4554
4. Zheng C, Zhou X, Cao H, Wang G, Liu Z (2014) Synthesis of porous graphene/activated carbon composite with high packing density and large specific surface area for supercapacitor electrode material. J Power Sources Thrower, Peter A., ed. Chemistry & physics of carbon. Vol. 25. CRC Press, 1996. 258:290–296
5. Harris P (1997) Chemistry and physics of carbon 13
6. Suárez D, Menéndez JA, Fuente E, Montes-Morán MA (1999) Contribution of pyrone-type structures to carbon basicity: an ab initio study. Langmuir 15(11):3897–3904
7. Garten VA, Weiss DE (1957) A new interpretation of the acidic and basic structures in carbons. II. The chromene-carbonium ion couple in carbon. Aust J Chem 10(3):309–328
8. Montes-moran MA, Menendez JA, Fuente E, Suarez D (1998) Contribution of the basal planes to carbon basicity: an Ab initio study of the H_3O^+—π interaction in cluster models. J Phys Chem 102(29):5595–5601
9. Calvo EG, Rey-Raap N, Arenillas A, Menendez JA (2014) The effect of the carbon surface chemistry and electrolyte pH on the energy storage of supercapacitors. RSC Adv 4(61):32398–32404
10. Wang H, Casalongue HS, Liang Y, Dai H (2010) $Ni(OH)_2$ nanoplates grown on graphene as advanced electrochemical pseudocapacitor materials. J Am Chem Soc 132(21):7472–7477
11. Peng WC, Bin Wang S, Li XY (2016) Shape-controlled synthesis of one-dimensional α-MnO_2 nanocrystals for organic detection and pollutant degradation. Sep Purif Technol 163:15–22
12. Nagarajan N, Humadi H, Zhitomirsky I (2006) Cathodic electrodeposition of MnO_x films for electrochemical supercapacitors. Electrochim Acta 51(15):3039–3045

13. Jang JH, Kato A, Machida K, Naoi K (2006) Supercapacitor performance of hydrous ruthenium oxide electrodes prepared by electrophoretic deposition. J Electrochem Soc 153(2):A321–A328
14. Qu QT, Shi Y, Li LL, Guo WL, Wu YP, Zhang HP, Guan SY, Holze R (2009) $V_2O_5{\cdot}0.6H_2O$ nanoribbons as cathode material for asymmetric supercapacitor in K_2SO_4 solution. Electrochem Commun 11(6):1325–1328
15. Sarma B, Ray RS, Mohanty SK, Misra M (2014) Synergistic enhancement in the capacitance of nickel and cobalt based mixed oxide supercapacitor prepared by electrodeposition. Appl Surf Sci 300:29–36
16. Tang H, Wang J, Yin H, Zhao H, Wang D, Tang Z (2015) Growth of polypyrrole ultrathin films on MoS_2 monolayers as high-performance supercapacitor electrodes. Adv Mater 27(6):1117–1123

Chapter 4
Graphene Analogous Elemental van der Waals Structures

Oswaldo Sanchez, Joung Min Kim, and Ganesh Balasubramanian

4.1 Elemental Structure

Although the atomic arrangements of Group IV elements follow the hexagonal honeycomb structure of graphene, only the carbon-based material forms a perfect planar layer. The other elements construct buckled hexagonal structures, as seen in Fig. 4.1 [1]. In Fig. 4.1a it is possible to see the buckled structure arrangement, where some of the atoms in a unit cell demonstrate a planar separation. There are different possible lattice arrangements that appear from this buckling behavior. These structures are illustrated in Fig. 4.1b, where the planar separation, or buckling distance, is designated by δ. From the image, the "Flat" structure demonstrates the planar structure found with graphene. For the boat and washboard structures, it has been found that they are unstable and will converge to the flat structure, while the chair structure demonstrates a greater stability than even that of the flat structure [2]. Table 4.1 below contains the lattice and buckling parameters for select Group IV elemental sheets.

The original version of this chapter was revised. An erratum to this chapter can be found at DOI 10.1007/978-3-319-64717-3_7

O. Sanchez (✉) • J.M. Kim
Department of Mechanical Engineering, Iowa State University,
Black Engineering Building, Ames, IA 50011-2030, USA
e-mail: osanchez@iastate.edu

G. Balasubramanian
Lehigh University, Bethlehem, Pennsylvania, USA

G. Balasubramanian (ed.), *Advances in Nanomaterials*,
DOI 10.1007/978-3-319-64717-3_4

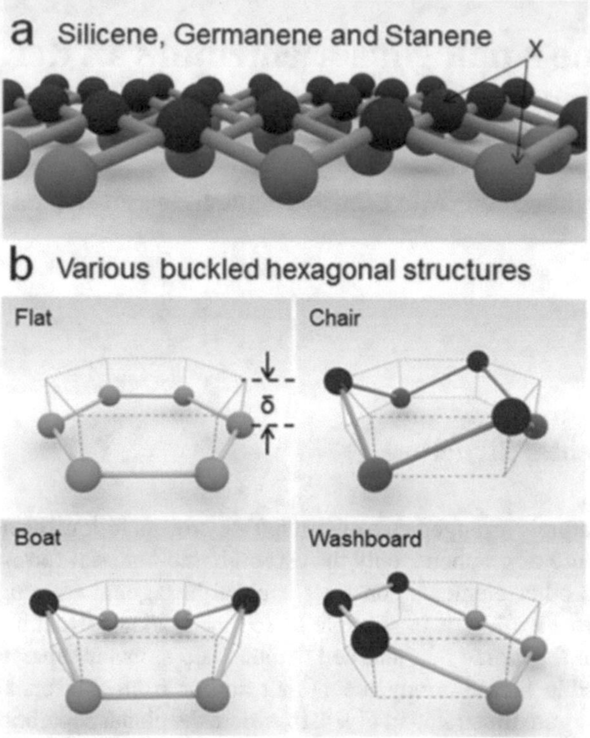

Fig. 4.1 (**a**) Buckled hexagonal crystal structure of 2D elemental sheets (X = Si, Ge, and Sn). Darker shaded atoms are at a slightly higher horizontal plane than the lighter shaded atoms. (**b**) Various hexagonal buckled structures of X. The buckling parameter δ is the vertical distance between the two planes of X atoms

Table 4.1 Structural and elemental features

	C	Si	Ge	Sn
Lattice constant a (nm)	0.2468	0.3868	0.4060	0.4673
Bond length d (nm)	0.1425	0.2233	0.2344	0.2698
Buckling parameter δ (nm)	0	0.045	0.069	0.085
Effective electron mass $_M$* (m_0)	0	0.001	0.007	0.029
Fermi velocity of carriers V_F (10^6 ms^{-1})	1.01	0.65	0.62	0.55
Energy gap E_g (meV)	0.02	1.9	33	101

Electronic quantities for Group IV elements are derived from hybrid exchange-correlation functional HSE06 calculations with the inclusion of spin orbit coupling [Table [1]] [Data [2]]

4.2 Silicene

The discovery of graphene and the tremendous advancements in this field of research have fueled the effort of searching for similar two-dimensional materials composed of Group IV elements, especially with silicon, via theoretical and experimental approaches.

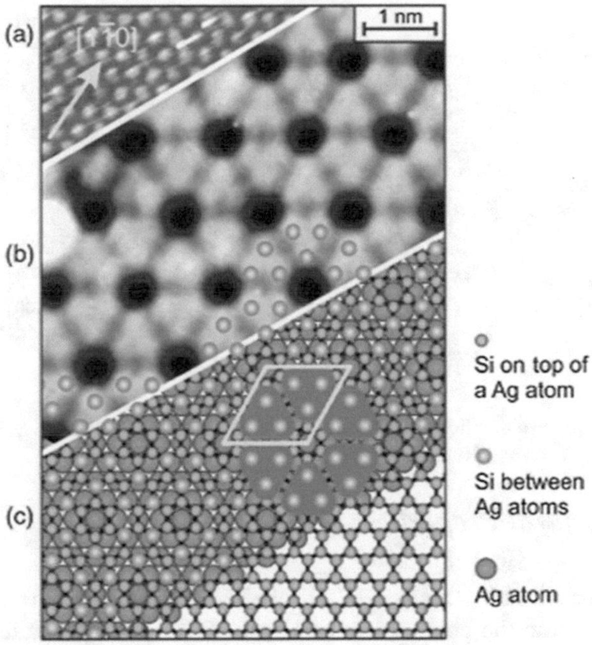

Fig. 4.2 Silicene on Ag (111) surface: Filled state STM images of (**a**) Ag (100) surface and (**b**) silicene on Ag. (**c**) Schematic overlay describing the observed STM images in (**a**) and (**b**) [taken from [1] which reproduced it from [3]]

4.2.1 Synthesis

One of the most fundamental but also greatest challenges comes with finding the proper way of synthesizing these 2D materials. Silicene does not seem to exist in nature, nor is there a solid phase as there is with graphene [3]. The synthesis of graphene has been possible via exfoliation of bulk graphite due to weak interlayer interactions [4]. Unfortunately, this is not the case for silicon, and for this reason, exfoliation methods utilized in synthesizing single-layer graphite could not generate pure 2D silicon layers [3, 4]. To get 2D silicon layers, researchers must consider the growth or synthesis of silicon with more sophisticated methods [3]. Among the methods, the most widely spread one is to deposit silicon on metal surfaces which do not interact strongly with the Si atoms or form compounds [3]. Many have successfully synthesized Silicene on Ag (111) via epitaxial growth [3, 5, 6]. This is particularly effective due to the interactions, or lack thereof, between Ag and Si. Davila [7] describes that Ag and Si "form atomically abrupt surfaces without intermixing". Figure 4.2 demonstrates a scanning tunneling microscopic (STM) images of a buckled silicene layer on Ag (111) as synthesized by Vogt et al. [3]. There has also been success in synthesizing buckled silicene on Ir (111) [4].

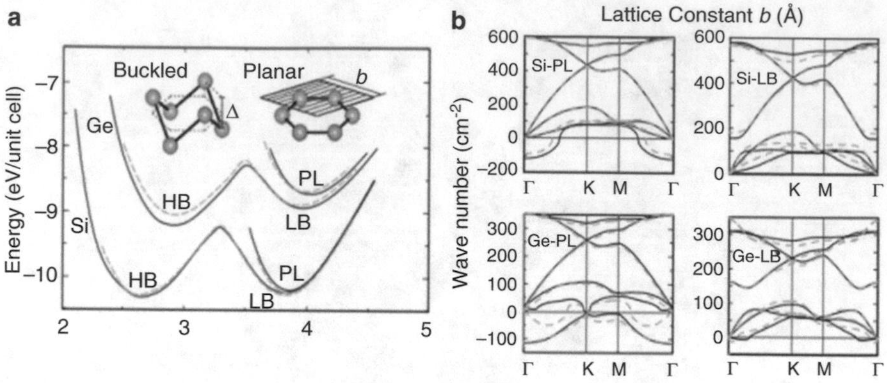

Fig. 4.3 Band structure of silicene and of Si (111) obtained from TB models compared to the ab initio results from Yang and Ni [11, 12]

4.2.2 Structural Properties

The planar honeycomb structure in silicene has an imaginary frequency in the Brillouin zone from the phonon mode analysis. During the structure optimization on a 2 × 2 supercell, there is a tendency to make a cluster in the high-buckled structure. For all kinds of silicene structures, the variation of binding energy remains constant. The surface of silicene is very reactive because of its weak interatomic bonds [8]. This makes silicene readily absorb chemical species, which forms chemical bonds with silicene [9]. Because of this, modification is adopted for the surface of silicene with transition metals. For example, with the existence of Ti and Ta in puckered silicene, the material becomes a planar structure while $NbSi_2$ shows the largest mechanical stiffness. For reducing reactiveness from the surface of silicene, doping is also used as an alternative. The buckling in silicene is largely influenced by the carrier concentration [8] (Fig. 4.3).

4.2.3 Electronic Properties

Similar to graphene, the π bands of silicene are also not connected to the bands because of its planar and orbital symmetries [10]. As far as bands are concerned, the crossings between π bands and σ bands, which occur in graphene, do not occur in silicene due to the lowered down valence bands in silicene [11]. The π band maintains its form as in graphene, however, when the π* band approaches Γ from the KΓ and the MΓ directions, the band changes [11]. Compared to graphene, the electrons move slower in silicene. The longer atomic distance makes for weaker π bonds in silicene compared to that in graphene because π bonds usually involve the conduction property in each material. It can be shown from Fig. 4.4 that the property is

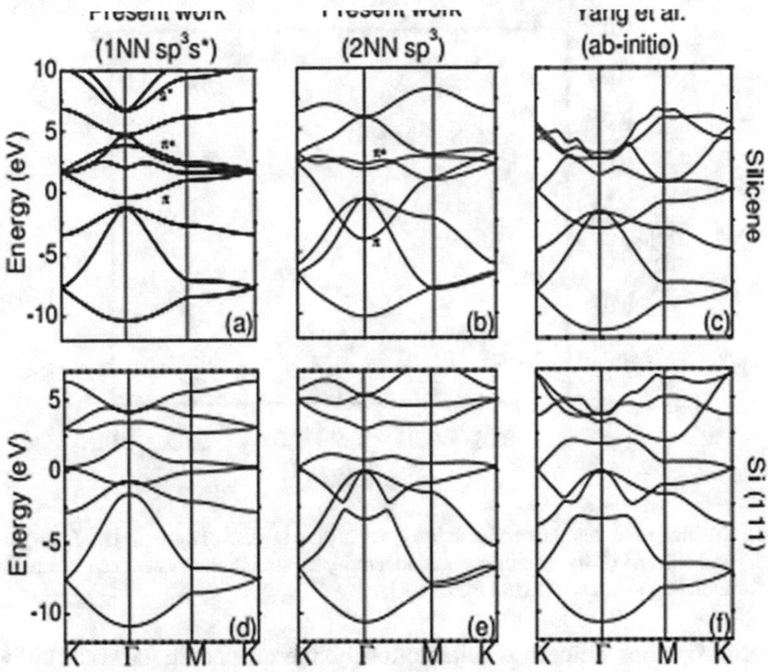

Fig. 4.4 (**a**) The variation of binding energy as a function of lattice constant for PL, LB, and HB honeycomb silicene and germanene and (**b**) the band structures of PL and LB silicene and germanene [13]

symmetry at K point. The π bands are weakened at K. Electrons around the K point should behave as Dirac massless fermions due to the presence of the Dirac cone in both structures. Compared to the fermi velocities of graphene (10^6 m/s), fermion velocities in Si (111) (10^4 m/s) and silicene (10^5 m/s) are slower. This is because the π interaction is weak in Si(111) and silicene [11].

4.2.4 Thermal Properties

From the previous studies, it is well-known that graphene has ultra-high thermal conductivities of 3000–5000 WmK^{-1} [14]. Compared to it, the recent studies on silicene show that the in-plane thermal conductivity of silicene at room temperature is in the range of 20–60 WmK^{-1}, which is almost 20% of that of bulk silicon, as extrapolated from the linear relation in Fig. 4.5 [15]. Comparably low thermal conductivity enables silicene to be more suitable to the purpose of thermoelectric figure. The large reduction in thermal conductivity of silicene compared to bulk silicon could be attributed to the increased phonon-surface scattering in low-dimensional semiconducting nanostructures. From these findings with excellent electric

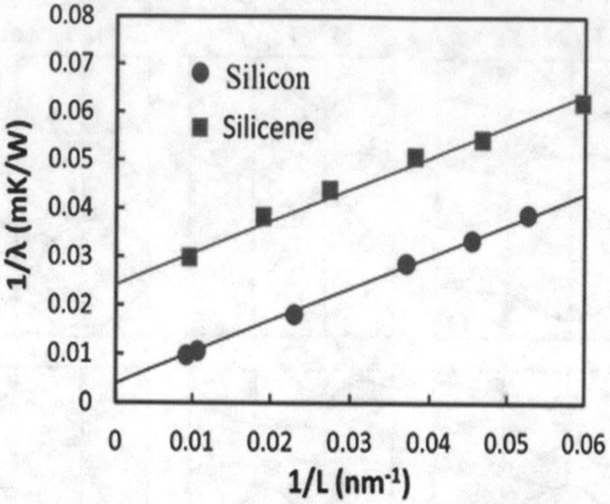

Fig. 4.5 The inverse of thermal conductivity, λ, versus the inverse of system size, L, for bulk silicon (^{28}Si) and silicene (^{28}Si). The thermal conductivity of the infinite system can be obtained by linear extrapolating to $1/L = 0$ [figure from [16]]

transport properties, silicene is suitable for the thermoelectric materials utilized in power generators or refrigeration applications.

The thermal conductivity of silicene is largely affected by the strain. With further increasing strain, the thermal conductivity starts to decrease. To explain this concept, at tensile strain of 0.12, the thermal conductivity drops more than 30% compared to the strain-free value, as seen in Fig. 4.6 [16]. For graphene, the thermal conductivity of graphene shows the state of decreasing all the way. However, the effect of strain in silicene shows small increase at the beginning phase [16]. This behavior of silicene could be attributed to the initial buckled configuration. The buckled configuration would be less buckled at small tensile strains, and this is because of bond rotation. As a result, we can get in-plane stiffness and an increase in the thermal conductivity in silicene [16]. Furthermore, it is also known that defects influence the thermal conductivity, from the studies on graphene. Among the defects, vacancy defects are quite unavoidable in 2D materials during growth and processing, and those defects are usually led by stress, irradiation, and sublimation [17]. Vacancy defects are lattice sites that in, a perfect crystal, would be occupied, but instead remain vacant. They not only affect electronic properties significantly, but also cause lattice vibrations localized around the defects. The localized vibration means that phonon thermal conduction will be reduced in both graphene and silicene, but similar research is still rarely performed in the field of silicene [17]. Increasing the concentration and size of the vacancy defects significantly reduces the phonon thermal conductivity of silicon nanosheets. In addition, not only the values of thermal conductivity, but also its anisotropy is influenced by the edge shape of vacancy clusters. This is considered important because of the chiral-angle tailoring of the thermal conductivity of silicene sheets [17]. Meanwhile, Isotope doping provides an efficient method to tune the

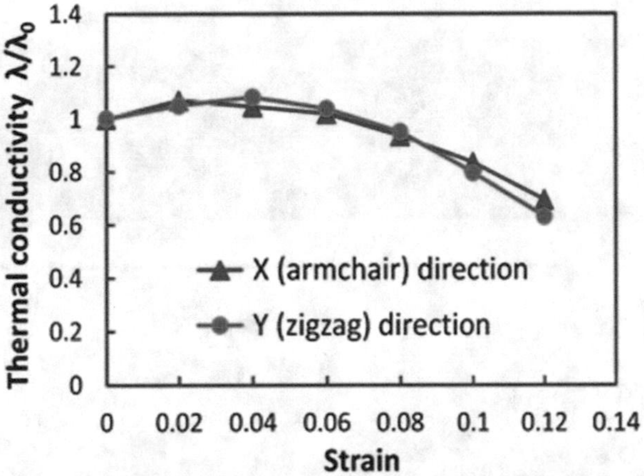

Fig. 4.6 Normalized thermal conductivity (λ/λ_0) of silicene as a function of tensile strain $(L-L_0)/L_0$ in the X (armchair) and Y (zigzag) directions. The thermal conductivities increase at small strains and decrease at large strains [figure from [16]]

thermal conductivity of nanomaterials. To get a result, Isotope doping is simulated by MD simulations. According to the simulation, the thermal conductivity of graphene and silicon nanowires can be dramatically reduced even at a low doping percentage. Moreover, ultra-low thermal conductivity can be achieved if the dopants are arranged into a superlattice structure, as in Fig. 4.7. Isotope doping has the advantage that it would not affect the electronic properties of the nanostructures since all the isotopic atoms have the same electronic structure. Researchers indicate that the bigger the mass difference between isotope atoms, the larger the reduction in thermal conductivity [16]. The maximum reduction in thermal conductivity is counted for 10 and 23% for Si and Si doping, each [16]. In addition, the graph of this result shows somewhat U-shaped change as the thermal conductivity correlated with the concentration of doping atoms. The thermal conductivity decreases initially to a minimum and then increases as the doping concentration changes from 0 to 100% [16]. The minimum of the thermal conductivity occurs at the doping concentration of around 50%. The randomly doped atoms can be considered as distributed impurities in the pure Si silicene lattice [16]. Those impurities cause the phonon scattering and localization of phonon modes, thus reducing the phonon group velocity. Therefore, the thermal conductivity decreases with increasing doping concentration. However, when the doping concentration is above 50%, the doped atoms become the main part of the lattice structure and the Si atoms become the impurities, as seen in Fig. 4.8 [16]. As a result, the phonon scattering and phonon-modes localization reduce with increasing doping atoms. Consequently, the thermal conductivity increases with increasing doping concentration from 50 to 100% [16]. Moreover, ultra-low thermal conductivity can be achieved if the dopants are arranged into a superlattice structure. Isotope doping has the advantage that it would not affect the electronic properties of the nanostructures since all the isotopic atoms have the same electronic structure. In addition to possible

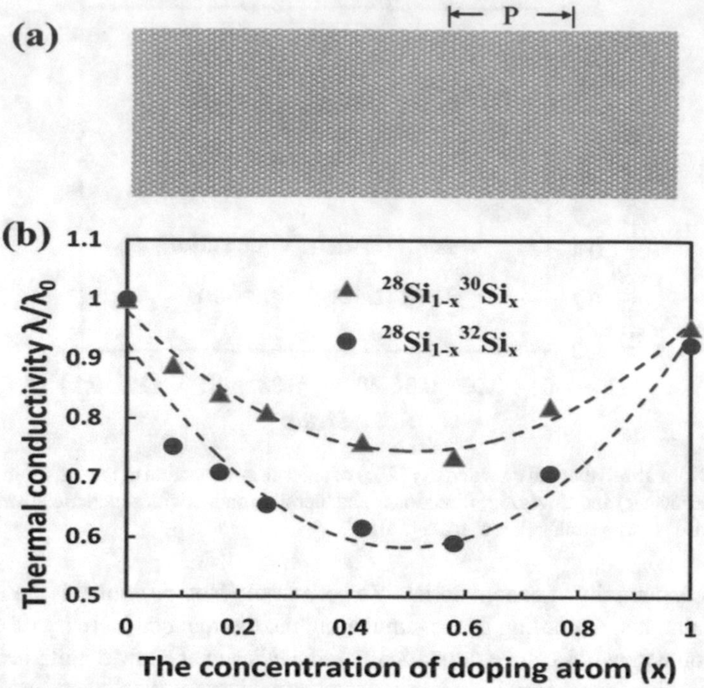

Fig. 4.7 (**a**) The atomic configuration of silicene with ordered doping (isotope superlattice), where the two different colors represent different isotopes. (**b**) Thermal conductivity of ordered doped silicene as a function of the percentage of doped isotope atoms [figure from [16]]

application for power generator and refrigerator, silicene can be used as a molecule sensor. Silicene can chemically absorb nitrogen-based molecules. NO_2 has the largest absorption energy of −1.12 and −1.53 eV per molecule for two absorption configurations, whereas the absorption energies of NO and NH_3 range from −0.46 to −0.60 eV per molecule [18]. The charge carrier concentrations of silicene are larger than that on graphene. These findings indicate that silicene is a potential candidate for a molecule sensor with high sensitivity for NH_3, NO, and NO_2 (Fig. 4.9).

4.3 Germanene

One of the novel graphene analogous materials gaining attention is germanene. This material is often connected and compared to silicene when reported in the literature. However, published experimental work on the material is comparably less than that on silicene. This is to be expected, as much of the current electronics technologies involve either silicon, germanium, or a combination of the two. Because of this, silicene and germanene offer opportunities for easy integration to current technologies, as opposed to graphene which would require significant modification before it

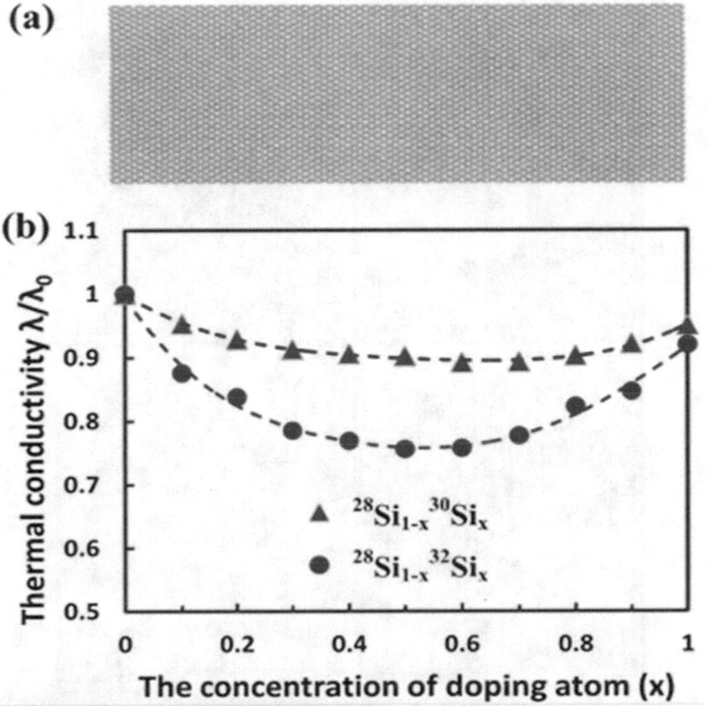

Fig. 4.8 (**a**) The atomic configuration of silicene with random doping, where the two different colors represent different isotopes. (**b**) Thermal conductivity of randomly doped silicene as a function of the percentage of doped isotope atoms [figure from [16]]

could be successfully implemented [1]. This, in turn, brings great appeal to investigations on properties and behaviors of silicene and germanene.

4.3.1 Synthesis

Germanene synthesis is even more novel than that of silicene. While Ag has been established as the common substrate for silicene synthesis, the use of Ag would not be plausible for germanene synthesis. At room temperature, Ge reactions with Ag form an Ag_2Ge alloy [22]. Because of this phenomenon, alternative substrates have been investigated and discovered. Davila et al. [7] hypothesized that the answer would lie in finding a material that demonstrated a similar behavior with Ge, as that of Ag with Si (with no intermixing); the material that was found to fit these criteria is Au (111) [7]. Once Au was identified as a suitable substrate, the same process of dry epitaxial growth used for the silicene growth was implemented to grow germanene on the gold substrate. As Davila et al. [7] expected, the germanene growth was found

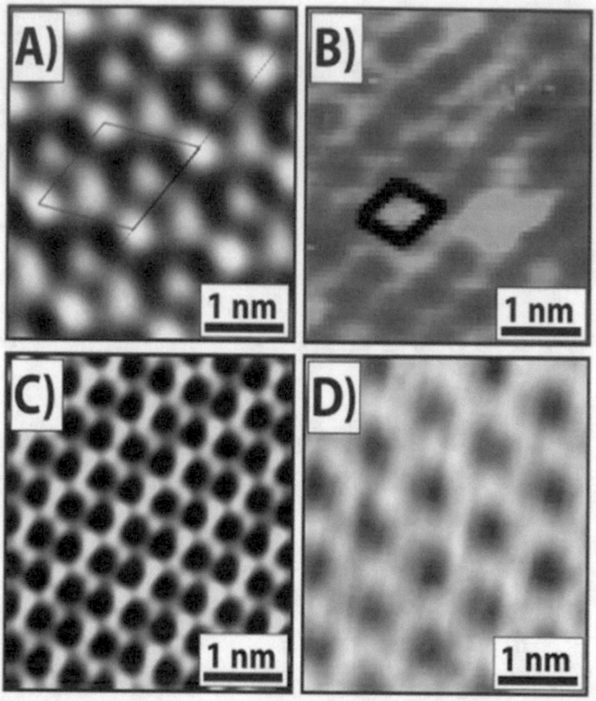

Fig. 4.9 STM images of germanene sheets grown by several research groups. For comparison, all the images have the same size of 4 nm × 4 nm. (**a**) STM image of germanene $\sqrt{19} \times \sqrt{19}$ superstructure on Pt (111). ($V = 1$ V and $I = 0.05$ nA) [19], (**b**) STM image of the germanene $\sqrt{3} \times \sqrt{3}$ superstructure on Au(111) ($V = -1.12$ V and $I = 1.58$ nA; *the Au(111)* $\sqrt{7} \times \sqrt{7}$ unit cell is outlined in *black*) [7], (**c**) STM image of the germanene honeycomb layer on Ge_2Pt cluster ($V = -0.5$ V and $I = 0.2$ nA) [20], and (**d**) STM image of the germanene 3 × 3 superstructure on Al(111) ($V = -0.7$ V and $I = 0.3$ nA) [21] [figure from [8]]

to be comparable to the silicene formation on Ag substrates. Continuous germanene layers have also been successfully synthesized on Al (111) [21] and Pt (111) [19, 23].

Unfortunately, the current synthesis methods do not cater to free standing germanene [24]. Miro et al. [24] mention that germanane (GeH) does not need a substrate to be stable; however, as of now, the synthesis of single-atom germanene is limited to a few substrates [25].

4.3.2 Structural Properties

Currently, the structure stability, electronic and vibrational properties have been investigated via ab initio calculations [26, 27]. As increased interatomic distance, the bonding between atoms in germanene is significantly weaker than that in graphene. This makes less energy distributions between the bonding and antibonding orbitals. It affects the structure of germanene. Because the band structure is deduced from planar

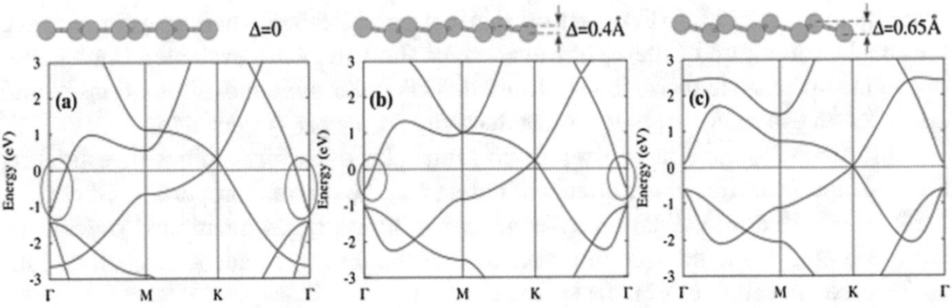

Fig. 4.10 Electronic band structure of germanene calculated using DFT for different values of the buckling Δ. Zero energy corresponds to the Fermi energy. *Blue circles* denote the antibonding band crossing the Fermi energy at low buckling values

germanene as can be seen from Fig. 4.10, the low lying antibonding is staying around Γ-point. It results in a finite density of state at the Fermi level. It is not preferable for its energy, so that achieving a buckled structure by expanding buckling (the vertical separation between two sub-lattices) of the low point group is highly recommended. The vertical distance between two sub-lattices is decided by a balance between the electronic and elastic energies. For free standing germanene, the buckling is ranged from 0.64 to 0.74 Å, and even the value becomes bigger than 2 Å at the total energy landscape. This structural property can be altered by interactions with other substrates. Like graphene, there is also the opening of a band gap induced by the symmetry in sub-lattice of supported germanene [25]. Matusalem et al. [26] found that a lower stability is obtained when in the graphene-like arrangement, and it becomes more stable in the honeycomb dumbbell arrangement. Acun et al. [25] reported DFT calculations on the buckled honeycomb structure of germanene and found it to be a 2D Dirac fermion system (this has yet to be validated by experiment). This leads to the hypotheses that the quantum spin Hall effect would be present at accessible temperatures [25]. This poses graphene as a promising 2D topological insulator [28].

4.3.3 Electronic Properties

Germanene possesses a semi-metallic band alignment and is predicted to possess similar electronic properties as silicene, including massless Dirac fermions [29, 30].

4.3.4 Thermal Properties

Currently, there are no available literatures focusing solely on thermal conductivities. This can be expected, as the main use of germanium, and silicon, applications involve thermoelectric properties. These studies are not normally found with graphene because the high thermal conductivity implies poor thermoelectric properties. Yang et al. [31] report on thermal conductance when investigating the thermoelectric figure of merit, however, since the values are evaluated implementing ab initio the system size is very limited. Molecular dynamics simulations would allow for an evaluation of thermal conductivity for larger system sizes.

Unfortunately, there are currently no molecular dynamics studies investigating thermal conductivities or the effects of defects on the thermal properties. Of course, these results require validation by experiment, however, as mentioned previously, current synthesis methods do not cater to free standing germanene so such investigations are unavailable (Fig. 4.11).

4.4 Stanene

Another Group IV graphene analogous material gaining popularity is stanene. An appealing characteristic of stanene stems from the fact that bulk tin is metallic, so there is an interest in exploiting those favorable electronic properties [14]. One of the appealing characteristics at this time is the potential for stanene to be established as a topological insulator [15, 33]. Essentially, this means good electrical conduction with minimal energy loss due to waste heat. This opens the potential for implementation of this material in electrical circuits.

4.4.1 Synthesis

The novelest synthesis discussed in this paper is that of monolayer tin. Unlike silicene and germanene, there are very few reports of successful synthesis of stanene. While the synthesis of 2D stanene has presented a challenge, there is a report of atomically thick free-standing few-layer stanene (FLS) that are characterized optically with UV–Vis absorption [34]. Also, Zhu et al. [32] present successful fabrication of stanene with molecular beam epitaxy (MBE) on Bi_2Te_3 (111). This success opens the possibility of experimentally investigating and validating the current theoretical and computational models that have been developed for stanene (Fig. 4.12).

4.4.2 Property Analysis

Due to the novelty of stanene, the available data are limited when it comes to experimental information. Since stanene synthesis is still an area of open investigations, the availability of samples to perform experiments is essentially nonexistent. Due to this, much of the available information has resulted from theoretical or computational analysis. Even then, since the focus has been mainly on silicene, and to a lesser extent, germanene, the knowledge available involving stanene is sparse at the moment. There

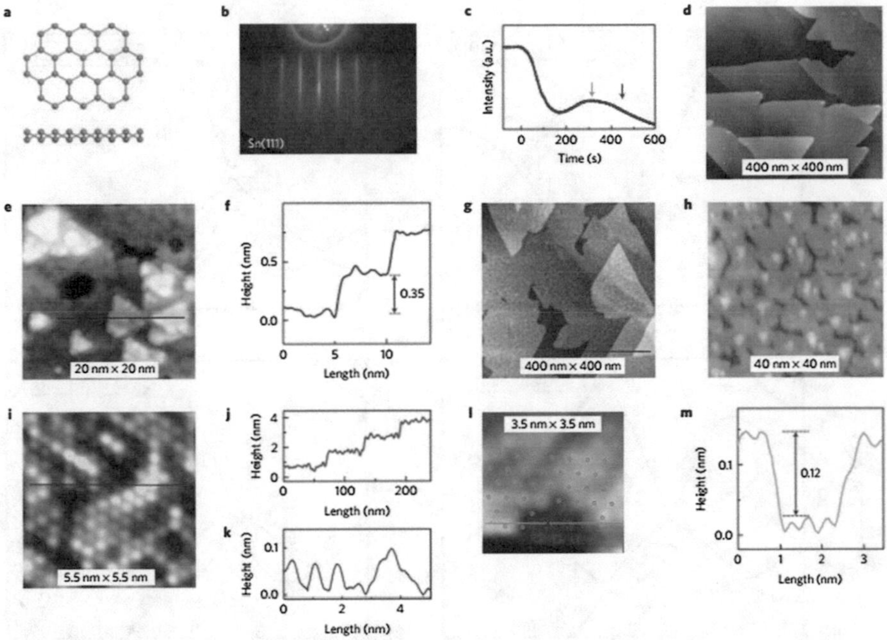

Fig. 4.11 Atomic structures of stanene on Bi_2Te_3. (**a**) Top view (*upper*) and side view (*lower*) of the crystal structure of stanene. (**b**) RHEED pattern of stanene film. (**c**) RHEED intensity as a function of growth time. The *blue arrow* marks the deposition time for stanene. (**d**, **e**) STM topography of Bi_2Te_3 (111) (**d**) and Sn films of more than single biatomic layer coverage (**e**). The corresponding deposition time is marked by the *black arrow* in (**c**). (**f**) Height line profile in (**e**). (**g**) Large-scale STM topography of stanene film. (**h**) Zoom-in STM image of stanene. (**i**) Atomically resolved STM image of stanene. (**j**, **k**) Height line profiles in (**g**) and (**i**). (**l**) Atomically resolved STM image of top and bottom atomic layers of stanene. *Blue dots* mark the lattice of the top Sn atoms. *Red dots* mark the lattice of the bottom Sn atoms. The two lattices do not coincide. (**m**) Height line profile in (**l**) [figure taken from [32]]

are reports available involving first principle, density functional theory, and Boltzmann transport equation calculations to analyze the thermal and mechanical properties of stanene [36, 37]. There are even some who are expanding into investigating the tuning of material properties of stanene. Garg et al. [38] performed DFT calculations to investigate the band gap opening in stanene by patterned B-N doping and then implemented MD simulations to confirm the stability of the structure.

4.4.3 Structural Properties

van den Broek et al. [14] presented first-principle DFT calculations to investigate the structural, mechanical, and electrical properties of 2D hexagonal tin. First-principle molecular dynamics calculations determined the monolayer to be thermally stable at temperatures up to 700 K. Mojumder et al. implemented MD

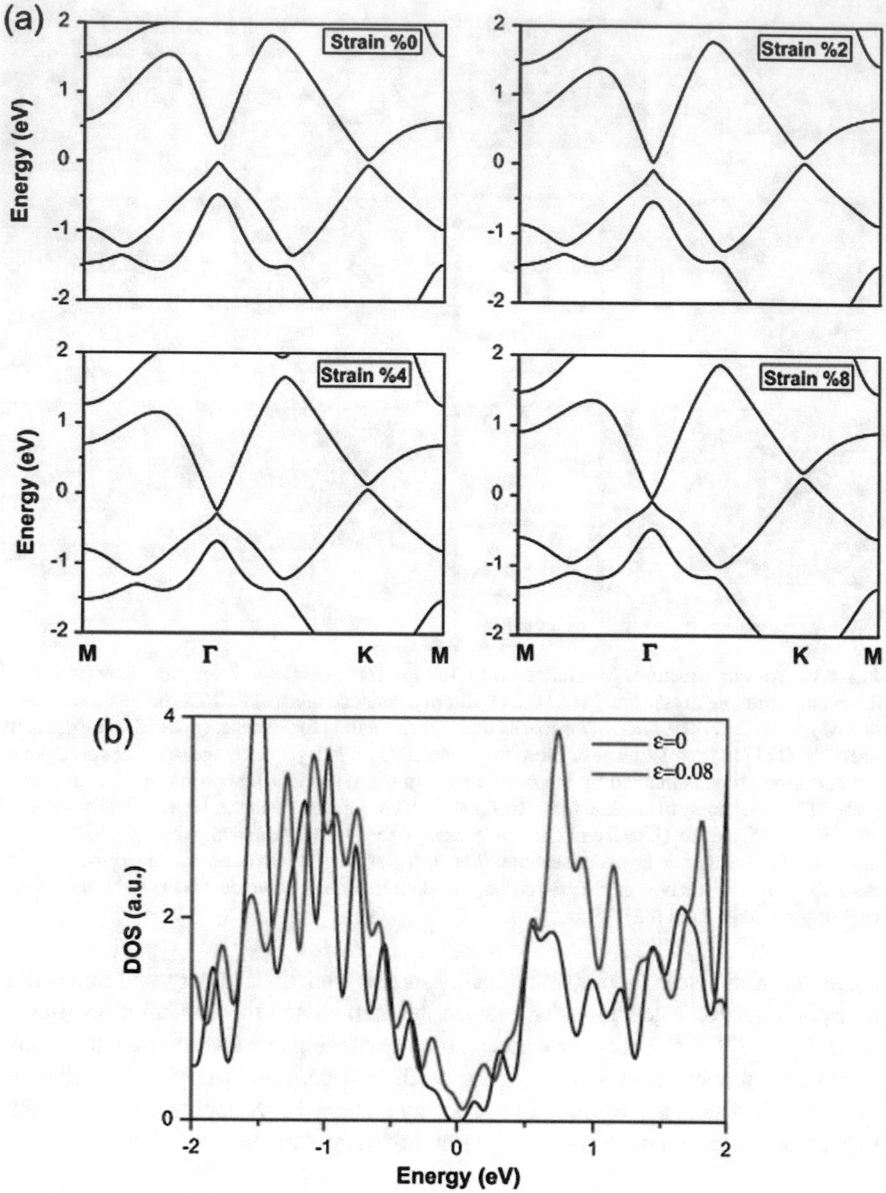

Fig. 4.12 (**a**) Band structure of stanene and (**b**) DOS for different values of applied strain in the presence of SOC [figure taken from [35]]

simulations with the embedded atom model to analyze mechanical properties of stanene. The resulting investigation showed that increased temperature causes a reduction in the fracture strength and strain on stanene. It was also notable that uniaxial loading in the zigzag direction presented a higher fracture strength and strain than that of armchair direction loading, while no noticeable difference was found for biaxial loading [39].

4.4.4 Electronic Properties

Stanene and germanene are very similar and are often compared to silicene together. As was the case for germanene, stanene has a slightly metallic band alignment and is also predicted to possess electronic properties like those found in silicene [29, 30].

4.4.5 Thermal Properties

Another area of interest in these 2D materials involves the thermal properties. If there is any hope of establishing stanene as a legitimate option for real applications, the thermal properties must be investigated. While information is limited, there have been some significant studies published on the matter. Peng et al. [37] presented an analysis of phonon transport in stanene via first principles calculations and phonon Boltzmann transport equations to evaluate the material's thermal conductivity. In fact, this analysis coincides with the results obtained by Nissimagoudar et al. [36] and establishes stanene as the material with the lowest thermal conductivity among all the Group IV materials. The latter report also mentions the potential for thermal conductivity to be tuned by adjusting the sample size and applying rough surfaces on the edges.

At this time, the analysis on thermal conductivity remains limited to pure stanene. Currently, there appears to be a lack of information on the effects of defects on thermal conductivity. Also, there are currently no molecular dynamics simulations available that model the thermal conductivity in this material. This can be attributed to the fact that there has not been a parameterization of suitable potentials for MD simulations.

References

1. Balendhran S et al (2015) Elemental analogues of graphene: silicene, germanene, stanene, and phosphorene. Small 11(6):640–652
2. Matthes L, Pulci O, Bechstedt F (2013) Massive Dirac quasiparticles in the optical absorbance of graphene, silicene, germanene, and tinene. J Phys Condens Matter 25(39)

3. Vogt P et al (2012) Silicene: compelling experimental evidence for graphene like two-dimensional silicon. Phys Rev Lett 108(15)
4. Meng L et al (2013) Buckled silicene formation on Ir(111). Nano Lett 13(2):685–690
5. Lalmi B et al (2010) Epitaxial growth of a silicene sheet. Appl Phys Lett 97(22)
6. Lin CL et al (2012) Structure of silicene grown on Ag(111). Appl Phys Express 5(4)
7. Davila ME et al (2014) Germanene: a novel two-dimensional germanium allotrope akin to graphene and silicene. New J Phys 16
8. Jose D, Datta A (2014) Structures and chemical properties of silicene: unlike graphene. Acc Chem Res 47(2):593–602
9. Kara A et al (2012) A review on silicene-new candidate for electronics (vol 67, pg 1, 2012). Surf Sci Rep 67(5):141–141
10. Saito R, Dresselhaus G, Dresselhaus MS (1998) Physical properties of carbon nanotubes, vol 35. World Scientific
11. Guzman-Verri GG and Voon LCLY (2007) Electronic structure of silicon-based nanostructures. Phys Rev B 76(7)
12. Yang X, Ni J (2005) Electronic properties of single-walled silicon nanotubes compared to carbon nanotubes. Phys Rev B
13. Cahangirov S et al (2009) Two-and one-dimensional honeycomb structures of silicon and germanium. Phys Rev Lett 102(23)
14. van den Broek B et al (2014) Two-dimensional hexagonal tin: ab initio geometry, stability, electronic structure and functionalization. 2D Materials 1(2)
15. Tang PZ et al (2014) Stable two-dimensional dumbbell stanene: a quantum spin hall insulator. Phys Rev B 90(12)
16. Pei QX et al (2013) Tuning the thermal conductivity of silicene with tensile strain and isotopic doping: a molecular dynamics study. J Appl Phys 114(3)
17. Li HP, Zhang RQ (2012) Vacancy-defect-induced diminution of thermal conductivity in silicene. Epl 99(3)
18. Hu W et al (2014) Silicene as a highly sensitive molecule sensor for NH_3, NO and NO_2. Phys Chem Chem Phys 16(15):6957–6962
19. Li LF et al (2014) Buckled germanene formation on Pt(111). Adv Mater 26(28):4820–4824
20. Bampoulis P et al (2014) Germanene termination of Ge_2Pt crystals on Ge(110). J Phys Condens Matter 26(44)
21. Derivaz M et al (2015) Continuous germanene layer on Al(111). Nano Lett 15(4):2510–2516
22. Oughaddou H et al (2000) Ge/Ag(111) semiconductor-on-metal growth: formation of an Ag_2Ge surface alloy. Phys Rev B 62(24):16653–16656
23. Svec M et al (2014) Silicene versus two-dimensional ordered silicide: atomic and electronic structure of Si-(root 19x root 19)R23.4 degrees/Pt(111). Phys Rev B 89(20)
24. Miro P, Audiffred M, Heine T (2014) An atlas of two-dimensional materials. Chem Soc Rev 43(18):6537–6554
25. Acun A et al (2015) Germanene: the germanium analogue of graphene. J Phys Condens Matter 27(44)
26. Matusalem F et al (2015) Stability and electronic structure of two-dimensional allotropes of group-IV materials. Phys Rev B 92(4)
27. Roome NJ, Carey JD (2014) Beyond graphene: stable elemental monolayers of silicene and germanene. ACS Appl Mater Interfaces 6(10):7743–7750
28. Le Lay G et al (2015) Increasing the lego of 2D electronics materials: silicene and germanene, graphene's new synthetic cousins. Micro- and Nanotechnology Sensors, Systems, and Applications VII 9467
29. Houssa M et al (2010) Electronic properties of two-dimensional hexagonal germanium. Appl Phys Lett 96(8)
30. Lebegue S, Eriksson O (2009) Electronic structure of two-dimensional crystals from ab initio theory. Phys Rev B 79(11)

31. Yang K et al (2014) Thermoelectric properties of atomically thin silicene and germanene nanostructures. Phys Rev B 89(12)
32. Zhu FF et al (2015) Epitaxial growth of two-dimensional stanene. Nat Mater 14(10):1020–1025
33. Xu Y et al (2013) Large-gap quantum spin hall insulators in tin films. Phys Rev Lett 111(13)
34. Saxena S, Choudhary RP, Shukla S (2016) Stanene: atomically thick free-standing layer of 2D hexagonal tin. Sci Rep 6
35. Modarresi M et al (2015) Effect of external strain on electronic structure of stanene. Comput Mater Sci 101:164–167
36. Nissimagoudar AS, Sankeshwar NS (2014) Significant reduction of lattice thermal conductivity due to phonon confinement in graphene nanoribbons. Phys Rev B 89(23)
37. Peng B et al (2016) Low lattice thermal conductivity of stanene. Sci Rep 6:20225
38. Garg P, Choudhuri I, Pathak B (2017) Band gap opening in stanene induced by patterned BN doping. Phys Chem Chem Phys 19:3660–3669
39. Mojumder S, Al Amin A, Islam MM (2015) Mechanical properties of stanene under uniaxial and biaxial loading: a molecular dynamics study. J Appl Phys 118(12)

Part III
Nanomaterial Interfaces and Nanofluids

Chapter 5
Nanostructured Oxides: Cross-Sectional Scanning Probe Microscopy for Complex Oxide Interfaces

TeYu Chien

5.1 Background

Charge transportation in materials has drawn the attention of physicists since the electronic nature of materials was observed—even before the electron was discovered (in the year 1897 by J. J. Thomson). In a material, electrons dwell in an environment full of ions and electrons. The many-body nature of the electronic properties in materials has proven that the modeling of the electronic properties is a very difficult task. Surprisingly, the Drude model, in which the electrons in metals were modeled as an electron gas without interactions (free electron gas), has described electronic properties for simple metals very well [1], despite the many-body nature of electrons in metals. Soon after the discovery of the Pauli exclusion principle for electrons, Sommerfeld applied quantum mechanics and the Fermi-Dirac distribution for the free electron gas model. With the more accurate modification, the Sommerfeld theory resolved some puzzles that had been thrown out by the Drude model, such as the Wiedemann-Franz law [1]. However, it still ignored the many-body nature of the electrons in metals. A more complicated model is needed for materials with strong interactions between electrons and other degrees of freedom, such as spin, lattice, and orbital. Beyond metals, theories based on weak correlations and interactions could also describe semiconductors very well. Over decades, the semiconductor physics worked very well to describe and predict the physical properties and behaviors of modern electronic devices.

Recently, since materials with strong electron correlation and interactions become the materials of the interest, semiconductor physics faces the fate of re-examination [2]. Great varieties of functionalities have been observed in complex oxide materials. For examples, colossal magnetoresistance (CMR) [3], superconductivity (SC) [4], ferroelectricity (FE) [5–7], and multiferroics (MF) [8] have been

T. Chien (✉)
Department of Physics and Astronomy, University of Wyoming, Laramie, WY 82071, USA
e-mail: tchien@uwyo.edu

G. Balasubramanian (ed.), *Advances in Nanomaterials*,
DOI 10.1007/978-3-319-64717-3_5

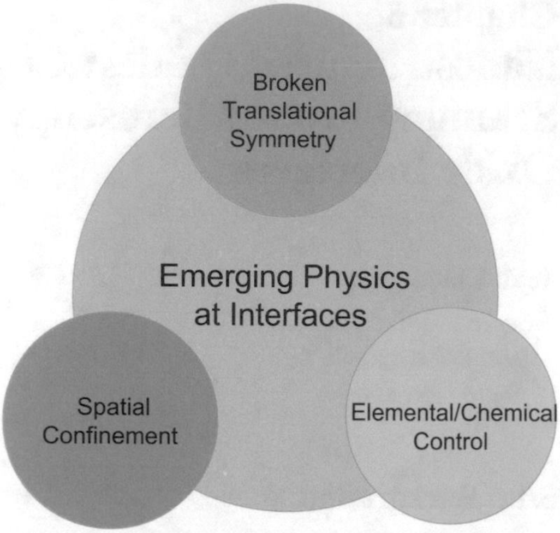

Fig. 5.1 Schematic of the relationships among the broken translational symmetry, spatial confinement, and elemental/chemical engineering with the emerging physics found at interfaces

observed in complex oxides. The goals of the thriving studies in complex oxide materials are focused on utilizing and manipulating the macroscopic properties, such as dielectric constant, superconductivity, and magnetism, of the materials for commercial use. One of the approaches is focused on understanding the aforementioned microscopic phenomena (CMR, SC, and MF) and further controlling/manipulating the macroscopic behaviors.

On the other hand, due to the spatially confined environment and broken translational symmetry, the electron correlations and interactions become very sophisticated in nanoscale materials. In two-dimensional (2D) environment, such as the surfaces of solid materials, structural [9], electronic [10], spin [11, 12], and orbital [13] reconstructions were observed as the consequence of the new lowest energy configurations with the presence of the spatial confinement and the broken translational symmetry. Not surprising, in addition to surfaces, these reconstructions (charge [2], spin [14], lattice [15], and orbital [16]) have also been observed at interfaces, where the spatial confinement and the broken symmetry still prevail. As depicted in Fig. 5.1, in addition to the spatial confinement and the broken translational symmetry, the elemental/chemical control at the interfaces acts as another important factor to affect the interfacial phenomena. Unlike the unchangeable spatial confinement and broken translational symmetry, the elemental/chemical control at interfaces provides an ideal way of engineering interfacial properties. A similar role is found for the interfaces of semiconductors, which have taken advantages of the interfacial properties in semiconductors for decades in the modern electronic devices. The search for next-generation electronic devices will depend on the understanding and further controlling the behavior at interfaces in next-generation materials, such as complex oxide materials.

There are two challenges to fully understand the physics at interfaces of complex oxides. The first challenge is the fabrications of the high-quality interfaces of complex oxides. Though the high-quality interfaces of semiconductors have been routinely synthesized, the control of the high-quality complex oxide interfaces is not trivial. The main issues were mainly coming from the complexity of the thermal dynamics and kinematics involving transition metals and oxygen during growth. With decades of efforts, nowadays, oxide molecular beam epitaxy (MBE) [17], and pulsed laser deposition (PLD) [18, 19] are the two most successful ways of synthesizing high-quality complex oxide thin films. With the high-quality thin film synthesis available, high-quality interfaces may be achieved.

The second challenge is the lack of appropriate tools for probing the buried interfaces with required spatial resolution as well as the capability of measuring physical properties simultaneously. For example, high-resolution cross-sectional transmission electron microscopy (HR-XTEM) could provide excellent spatial resolution, down to atomic or even subatomic resolution, but lack information about the electronic density of states (DOS) near Fermi energy. On the other hand, synchrotron x-ray magnetic circular dichroism (XMCD), synchrotron X-ray magnetic linear dichroism (XMLD), and electronic transport measurement could provide information on physical properties but are techniques averaging over a large measuring area. What is needed for the aforementioned novel interfacial phenomena is a tool with excellent spatial resolution as well as the ability to extract physical properties. Among the available tools, various types of scanning probe microscopy (SPM) are the ideal groups of techniques that meet these requirements. On one hand, SPM is known for high spatial resolution, ranging from tens of nm down to subatomic resolution. On the other hand, various types of SPM provide a wide array of capabilities of probing many different kinds of physical properties. For examples, scanning tunneling microscopy and spectroscopy (STM/S) provides the capability of probing local electronic DOS (LDOS) with atomic resolution in real space. Conducting atomic force microscopy (cAFM) is capable of probing local conductance with tens of nm resolution. With these available tools, however, the major challenge for using the SPM to study the interfaces is how to prepare samples for SPM to probe the buried interfaces. One of the approaches is to prepare the samples in cross-sectional geometry to expose the region of interests—interfaces—for SPM measurements. Here, we will briefly review the history of the recent developments of XSPM on complex oxide materials followed by many successful cases of XSPM measurements.

5.2 Complex Oxides and Complex Oxide Interfaces

One of the most studied families in complex oxides is the Ruddlesden-Popper (RP) series oxides, which have a common chemical formula as $AO(ABO_3)_n$. Four atomic structures of the RP series oxides are shown in Fig. 5.2. Basically, the structures of the RP series are composed of "n" layers of octahedron, BO_6, separated by a spacer, AO/AO stacking, in between. For examples, A_2BO_4 ($n = 1$) has single-layer octahedrons between two spacers; $A_3B_2O_7$ ($n = 2$) has double-layer corner-shared octahedrons between two spacers; and ABO_3 ($n = \infty$) has infinite corner sharing octahedrons

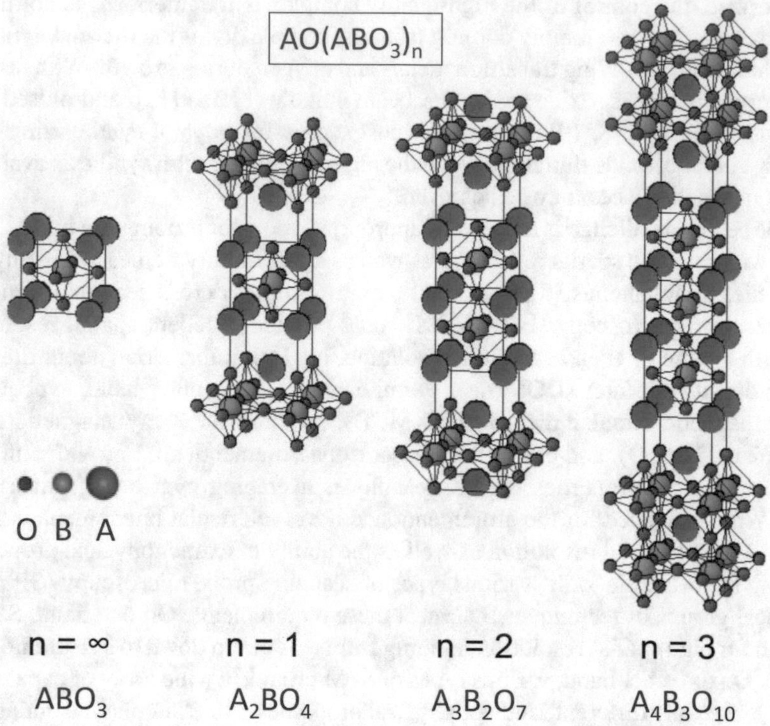

Fig. 5.2 Schematics of crystal structures of RP-series perovskite materials with $n = 1, 2, 3$, and ∞ in $AO(ABO_3)_n$

without the spacer. With the common structures, the RP series oxides could exhibit great varieties of functionalities by merely changing the elements A and B. For example, CMR effects in manganese-based perovskite oxides ($La_{1-x}Ca_xMnO_3$ or $La_{1-x}Sr_xMnO_3$) [3, 20], SC in copper-based oxides ($La_{2-x}Sr_xCuO_4$) [4], FE in titanium-based oxides [6], and MF in $BiFeO_3$ [8] were reported.

The great varieties of the functionalities found in the complex oxides are due to the highly coupled environment. To be clear, the interactions between electrons are referred as "electron correlation" or "correlation" in short; the interactions between electrons and other degrees of freedom are referred as "coupling". In materials with weak correlations, as mentioned above, Fermi gas model for describing the electronic properties is very successful. In Fermi gas model, the coupling between the electrons with other degrees of freedom, such as phonons and spin waves, are also neglected. In fact, for complex oxides, with the highly coupled and correlated environments, charge, orbital, spin and lattice degree of freedoms are mingled severely (Fig. 5.3). In complex oxides, weak correlation and coupling are no longer valid and further understanding, experimentally and theoretically, is required for complex oxide materials prediction. Some theoretical [2, 21, 22] and experimental [23–28] efforts have been embarked toward re-evaluating and understanding the strong interactions and coupling in complex oxides microscopically.

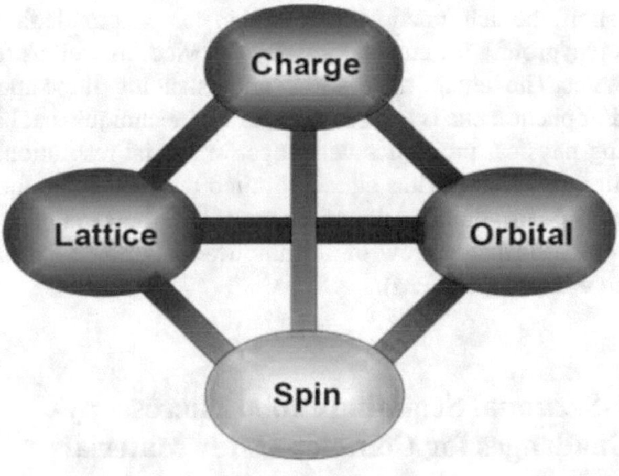

Fig. 5.3 Schematic describes the interplays among the four various degrees of freedom in complex oxide materials

The properties of the complex oxides could be controlled at least by, but not limited to, (1) element changing [29]; (2) element substitution/doping [20, 30–34]; (3) amount of oxygen vacancies [33–36]; (4) strain [37]; (5) temperature [29, 34]; and (6) radiation illumination [38]. For example, as mechanism (1), the conductivity of the $RENiO_3$ could be tuned by changing the rare earth (RE) elements [29], where $LaNiO_3$ is metallic at all temperatures studied [29]. On the other hand, the element substitution (mechanism (2)) could determine the magnetic phases of $La_{1-x}Ca_xMnO_3$ [20] and $La_{1-x}Sr_xMnO_3$ [30–32] and the superconducting temperature of $Bi_2Sr_{2-x}La_xCuO_6$ and of $La_{2-x}Sr_xCuO_4$ [33]. For mechanism (3), in complex oxides, the oxygen vacancies are generally considered as doping. The conductivity could induce with the presence of the oxygen vacancies in the otherwise insulating $SrTiO_3$ [34, 39]. Oxygen vacancies in $YBa_2Cu_3O_{7-\delta}$ (YBCO) also play a role as hole-doping to tune the charge carrier density hence the superconducting transition temperature [33]. It is worth noting that both element substitution (mechanism (2)) and oxygen vacancies (mechanism (3)) could be seen as a way of doping. The ferroelectricity could be induced by with the mechanism (4)—strain—in the otherwise ferromagnetic $EuTiO_3$ [37]. And finally, example of the mechanism (5) was reported as the two-dimensional electron liquid could be created on the bare $SrTiO_3$ surfaces by illuminating with ultraviolet irradiation [38].

When combining two different complex oxide materials together, extraordinary properties or phenomena could emerge at the interfaces. For example, two-dimensional electron gas was reported at the interfaces of two band insulators, $LaAlO_3/SrTiO_3$ [40], or Mott insulator and band insulator, $LaTiO_3/SrTiO_3$ [2]; Superconductivity was reported at $LaAlO_3/SrTiO_3$ interfaces [41] and at $La_2CuO_4/La_{1.55}Sr_{0.45}CuO_4$ interfaces [42]. The interplay between superconductor and ferromagnetic materials, such as $YBa_2Cu_3O_{7-\delta}/La_{2/3}Ca_{1/3}MnO_3$, shows the complex interactions between the cooper pairs and ferromagnetic materials near the interfaces [24]. In short, the rich functionality found in complex oxide families and their interfaces is a playground for engineering novel devices as well as for discovering interesting physics. The key for microscopic understanding of the above-mentioned properties and/or phenomena is to have a proper tool/technique that has the capability of extracting physical properties with superior spatial resolution. The superior spatial resolution requirement has already limited the tool/techniques to be somewhat "microscopy". In addition, the requirements of the capability of measuring physical properties further narrow the techniques/tools down to various types of scanning probe microscope (SPM).

5.3 Cross-Sectional Scanning Probe Microscopy and Challenges for Complex Oxide Materials

The key to use SPM for an interfacial study is to create the cross-sectional view of the interfaces. In general, issues on probing complex oxide interfaces in cross-sectional geometry using SPM could be considered in two aspects: (1) the tools (in SPM family); and (2) the methods for creating the cross-sectional view of the complex oxide interfaces. Each of the aspects puts limits on what could be measured and special set-ups might be necessary for successful studies.

First, the main differences among different types of SPMs are the spatial resolution, the extracted information, and the sample preparation requirements. For examples, scanning tunneling microscopy and spectroscopy (STM/S) has atomic or sometimes subatomic spatial resolution, but the samples are required to be conductive for the tunneled electrons being conducted away. The information obtained by STM/S is typically the topography and the electron local density of states (LDOS). On the other hand, conducting atomic force microscopy (cAFM) could be used for materials that are conductive, non-conductive or even mixed, but the spatial resolution is always limited to the tip apex size, typically in the order of 10 nm. The information obtained by cAFM is mainly the morphology and the local conductance. Electrochemical strain microscopy (ESM) measures the electrochemical reactivities with lateral spatial resolution in the order of 10 nm. STM, cAFM (including AFM) and ESM are the SPM techniques that have successful measurements on the complex oxide interfaces reported in the literature. Other types of SPM techniques potentially suitable for cross-sectional complex oxide measurements are: piezoresponse force microscopy (PFM), which may be used to study the piezoresponse at the complex oxide interfaces and/or domain walls; Kelvin probe force microscopy (KPFM) for visualizing the work function evolution across the interfaces of dissimilar complex oxide materials; magnetic force microscopy (MFM) for probing the magnetic domain across the interfaces of the complex oxides with magnetic interactions; scanning tunneling potentiometry (STP) for studying the equipotential profile across the complex oxide interfaces while an external bias was applied across the interfaces; and synchrotron X-ray scanning tunneling microscopy (SXSTM) [43–49] with elemental resolving power to study ion diffusion profile or composition in cross-sectional view.

Second, different methods to create the cross-sectional view of the interfaces for the SPM measurements have their own advantages and disadvantages. Traditionally, to prepare suitable surfaces for SPM studies, many methods have been used, including sputtering/annealing, polishing/annealing, and cleaving. Readers who are interested in these aspects of sample preparation please refer to reviews written by Bonnell et al. [50, 51]. Up to date, several successful methods have been reported for XSPM measurements: (1) fracturing; (2) polishing; (3) focused ion beam milling [23, 27, 52–55]. Each method has its own advantages/disadvantages as well as suitable SPM techniques. For example, fracturing could create contamination-free cross-sectional interfaces with a poor understanding of how the fracturing process affects the fractured results in atomic scale near the interfaces. Polishing is relatively easy to handle, however, contaminations and heat generated by the polishing procedure may be present. Focused ion beam milling could create a nicely controlled cross-sectional view of the interfaces, however, is more time-consuming and may have the intermediate level of contaminations.

Even for those successful methods of creating cross-sectional interfaces for SPM measurements, optimizing the preparation procedure is still under investigation. Take fracturing method as an example, the understanding of the fracturing/cleaving process of the materials of interests is important to precisely control the resulting fractured/cleaved topography, especially the regions near the interfaces. The fracturing process and results are determined by the fracturing dynamics and fracturing toughness. Microscopically, it has also been pointed out that the crack propagation is entirely determined by atomic-scale phenomena at the atomically sharp crack tips by breaking the inter-atomic bonds, one at a time at each point of the moving crack front [56]. In fact, the fracturing/cleaving properties of the complex oxides are relatively unexplored and could vary from one to another. For example, layered RP series materials will be cleaved at certain atomic layers (cleavage planes) and result in atomically flat surfaces when a shear force or tensile force applied on the samples along in-plane or out-of-plane directions, respectively. On the contrary, ABO_3 ($n = \infty$) materials do not have cleavage planes and will be "fractured", instead of "cleaved", with macroscopically rough surfaces, such as conchoidal fracturing morphology [57, 58], upon the application of a shear force. Interestingly, despite the macroscopic roughness seen in optical and electron microscopes [58], the existence of the atomic terraces with one unit cell or half unit cells terrace steps were revealed by STM measurements of $SrTiO_3$ [58–60]. Though the layered perovskites have cleavage planes, it is important to note that the cleavage planes are usually not aligned with the cross-sectional geometry, makes them still non-cleavable (though different manner) for XSPM studies.

The reported controllable fracturing process for perovskite complex oxide materials was done by first dicing the samples followed by in-situ fracturing prior to the STM/S measurements under ultrahigh vacuum (UHV) environment [24, 25, 58–62]. The sample thickness was either 1.0 mm or 0.5 mm in the studies done with Nb:STO [24, 25, 58–62] while the dimension of the samples were ~8 × 2 mm^2 or ~8 × 1 mm^2, respectively. The dicing depth could be precisely controlled to the range from 50 to 70% of the thickness by using a precision dicing saw. The diced

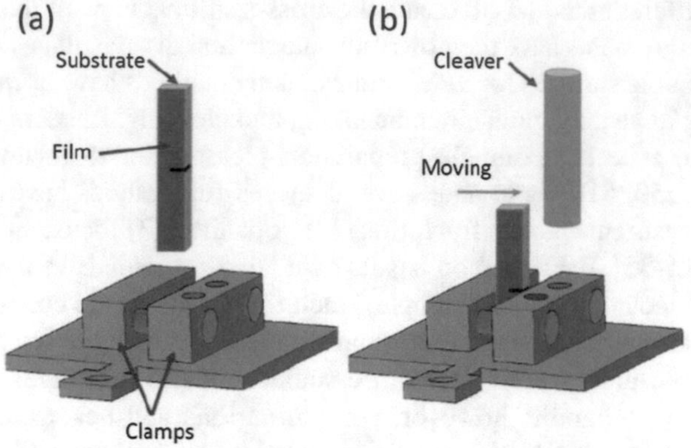

Fig. 5.4 Schematics of the sample-fracturing method for XSTM/S measurements preparation. (**a**) Before the fracturing with sample shape and relationship with the clamps on the sample holder. (**b**) The process of fracturing

samples are mounted mechanically and rigidly with a set of home-built clamps where the notch is aligned with the top surface of the clamps. Then the samples are fractured in-situ by moving the mounted samples against a cleaver in the UHV chamber, as shown in Fig. 5.4 [61, 62].

On the other hand, polishing and focused ion beam milling are the other methods reported to successfully create a cross-sectional view of the complex oxide interfaces [53–55, 63]. There are no detail studies about how the preparation parameters affect the resulting cross-sectional surfaces of the materials. In general, one expects contamination and/or heat damage for the samples prepared by the polishing method, while contamination and time consumption for the focused ion beam milling method. Further efforts are needed to gain insights on how to extend the use of these methods for XSPM measurements.

Hereafter, we will start to discuss what has been reported on complex oxide interfaces using above-mentioned preparation methods and their findings for pushing the understanding of complex oxide interface physics forward.

5.3.1 Controlled Fractured Surfaces of Nb-Doped $SrTiO_3$

5.3.1.1 Dual Terminations on the Fractured Nb-Doped $SrTiO_3$

One of the first reported fractured complex oxide materials probed by SPM is Nb-doped $SrTiO_3$ (Nb:STO) [58–60]. This study not only served as a testing trial for finding fracturing parameters for subsequent interfaces studied [23–25, 61, 62], but also revealed interesting fractured surfaces that deserve further investigation. This surprising observation was the well-ordered dual terminated (SrO and TiO_2 terminated surfaces) striped on the fractured Nb:STO surfaces [59, 64].

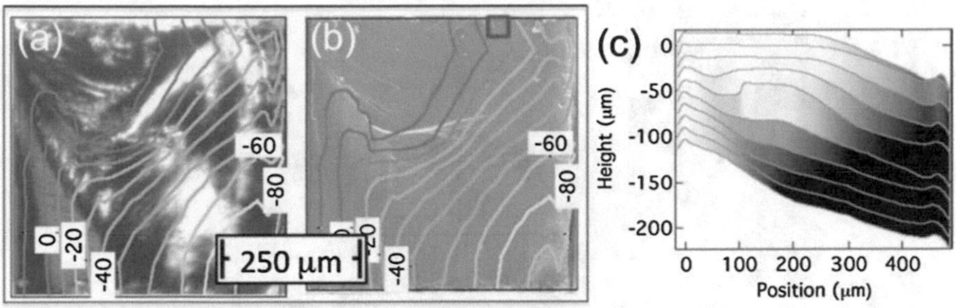

Fig. 5.5 (**a**) Optical microscope image; (**b**) SEM image; and (**c**) height profile of the fractured Nb:STO single crystal. The contour plot of the height profile was also plotted with the optical microscope and SEM images. Reproduced from J Vac Sci Technol B 28, C5A11 (2010), with the permission of AIP Publishing

As discussed above, perovskites with ABO_3 chemical formula do not have cleavage planes. For materials without cleavage planes, such as glass, it is known that the fractured morphology is composed of few distinct types: mirror, mist, hackle, and conchoidal lines [65]. These features are macroscopically visible by optical microscope or even by naked eyes, indicating the roughness of the fractured surface is at least in the order of few hundreds nanometer or even in micrometer scales. The question is: do the fractured surfaces of crystalline perovskites exhibit similar macroscopic morphological features, and if so, what are the microscopic morphological features imaged by STM? In order to answer this question, Chien et al. performed a survey of the fractured Nb-doped $SrTiO_3$ (Nb:STO) surfaces ranging from micrometer scale all the way down to nanometer scale, using various imaging techniques (optical microscopy (OM), scanning electron microscopy (SEM), and scanning tunneling microscopy (STM)) [58]. As shown in Fig. 5.5, by performing the controlled fracturing process at room temperature (RT), the fractured Nb:STO surfaces exhibit hundreds of micrometer roughness macroscopically, observed by profilometer, and the rough surfaces were clearly visible in optical and electron microscopes [58]. Conchoidal lines, mirror-like morphologies are observed in macroscopic scale (OM and SEM), which is similar to the fractured surfaces of glasses [65]. On the other hand, in nanometer scale as shown in Fig. 5.6, integer and half-integer unit-cell high steps were observed in STM images, indicating the mixed SrO and TiO_2 terminated surfaces [58]. In fact, the mixed terminated surfaces are expected, since there are no relative strong/weak bonded layers in the STO structure. As illustrated in Fig. 5.6h, no matter where the cracking appears in the crystal structure, the SrO and TiO_2 terminated surfaces will appear on either side of the fractured pieces [64]. The surprising findings are that the two types of terminations formed long range, ordered stripes, which are first reported by Guisinger et al. [59].

As illustrated in Fig. 5.7a, the fractured morphology of Nb:STO shows two sets of fracturing patterns. One is curved, which is similar to the fractured surfaces due to local stress-field (conchoidal fractures) [65]. The other one is straight, which is found to be perpendicular to [100] crystalline directions (fracturing direction).

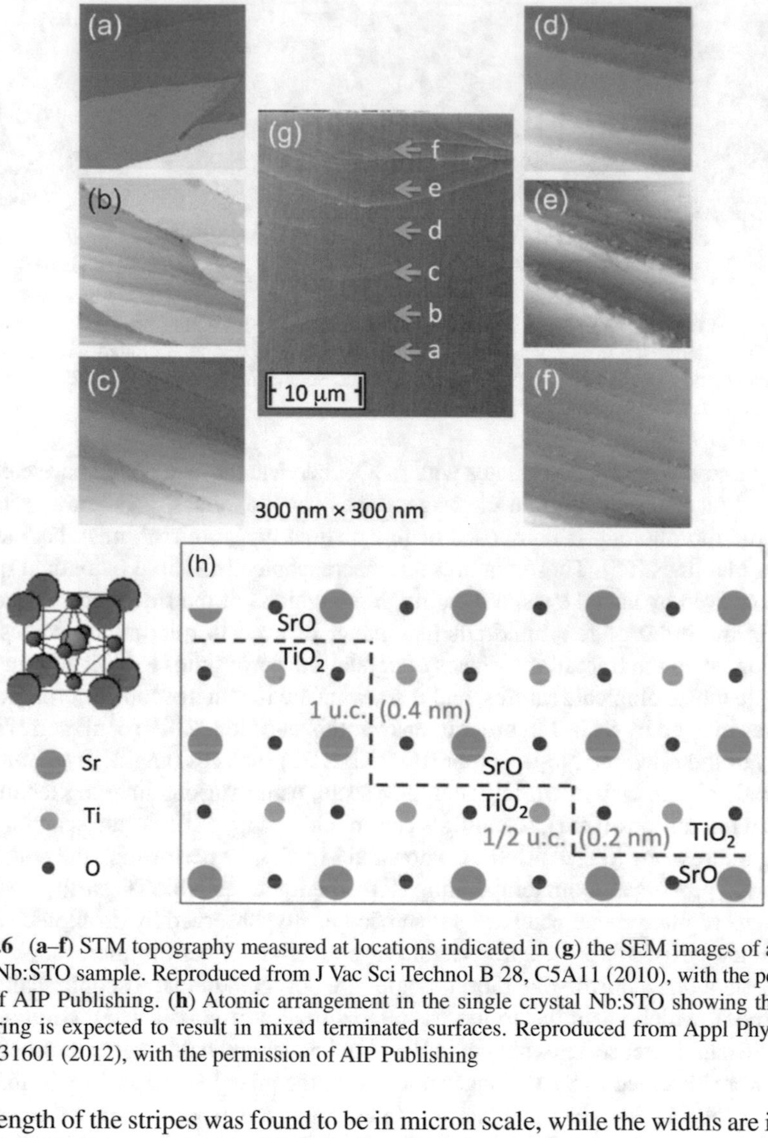

Fig. 5.6 (**a–f**) STM topography measured at locations indicated in (**g**) the SEM images of a fractured Nb:STO sample. Reproduced from J Vac Sci Technol B 28, C5A11 (2010), with the permission of AIP Publishing. (**h**) Atomic arrangement in the single crystal Nb:STO showing that the fracturing is expected to result in mixed terminated surfaces. Reproduced from Appl Phys Lett 100, 031601 (2012), with the permission of AIP Publishing

The length of the stripes was found to be in micron scale, while the widths are in the range of 10–20 nm scale [59, 64]. With the presence of these two types of fracturing features, one can find that the step heights change: when crossing the curved steps the step heights are integer unit-cell height; while when crossing those straight steps the step heights are half-integer unit-cell height, as seen in Fig. 5.7b. In other words, the surface terminations change from SrO to TiO_2 (or from TiO_2 to SrO) when crossing straight steps and remain the same while crossing curved steps. Furthermore, the two terminations exhibited distinct roughness: one is found be 0.2 nm (rough surface) and the other is 0.05 nm (smooth surface) [59]. The dual termination picture is further confirmed with d*I*/d*V* mapping and spectra measurements. As shown

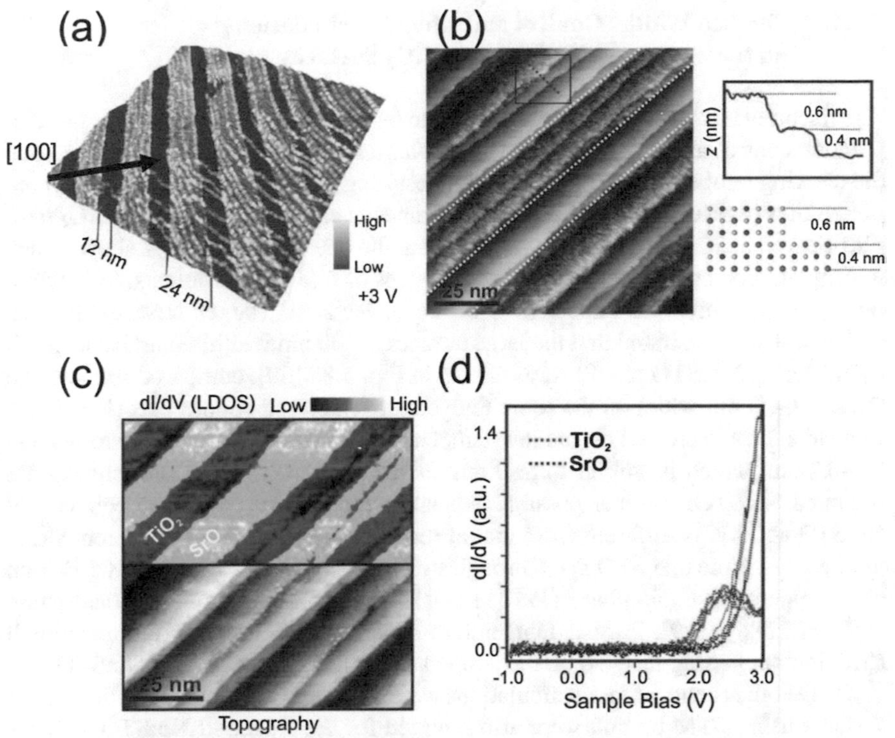

Fig. 5.7 (**a**) STM topography texture with d*I*/d*V* mapping color of the fractured Nb:STO surfaces. Reprinted with permission from ACS Nano 3, 4132 (2009). Copyright 2009 American Chemical Society. (**b**) STM topography showing two kinds of roughness on the fractured Nb:STO surfaces. The side view of the atomic arrangement along with the line profile showing the change of the types of roughness is accompanied by the change of half integer unit cell—meaning change of termination. (**c**) The d*I*/d*V* mapping along with the STM topography exhibited very straight termination boundaries. (**d**) The measured d*I*/d*V* point spectra on the two different types of termination showing distinct on-set bias in the conduction band minimum. This comparison of the d*I*/d*V* spectra to calculate PDOS of STO results in the assignment of the terminations for the observed topography. (**b–d**) Reprinted with permission from Adv Funct Mater 23, 2565 (2013). Copyright 2013 WILEY-VCH Verlag GmbH & Co

in Fig. 5.7c, the two different terminations exhibited d*I*/d*V* contrast in the mapping mode and clearly different d*I*/d*V* spectra (Fig. 5.7d). Note that the d*I*/d*V* signal is correlated to the electron local density of state (LDOS) and the STM is a surface-sensitive measurement. By comparing with density functional theory calculations [66, 67], the two terminations could be assigned as that the rough region is SrO and the smooth region is TiO_2 terminated surfaces [59]. The origin of the appearance of the straight fractured steps is still unclear. However, since it is straight relative to the crystalline directions, one possible explanation is the result of the “slip bands”, formed during the bending introduced by fracturing force. Further experiments are needed to confirm this hypothesis.

5.3.1.2 Terrace Widths Control and Atom Manipulations on the Fractured Nb-Doped $SrTiO_3$ Surfaces

The fractured Nb:STO surfaces were found to be controllable in certain ways—temperature control and electronic bias manipulations. First, it has been reported that the cleaving temperature could control the resulting cleaved surfaces for the layered perovskites (Sr_2RuO_4) [68]. In particular, Pennec et al. compared two Sr_2RuO_4 samples cleaved at 20 and 200 K and revealed that the one cleaved at 20 K shows much fewer point defects [68]. Considering step edges as one type of defects, lower fracturing temperatures are expected to create fewer steps (larger terrace). Indeed, Chien et al. demonstrated that the large terraces (~500 nm width) could be achieved by fracturing Nb:STO at ~50 K, as shown in Fig. 5.8a [60], compared to the small terraces (~20 nm wide) on the room temperature fractured counterpart (Fig. 5.7a). The large terrace created on the low-temperature fractured Nb:STO has roughness of ~0.2 nm, which is similar to SrO terminated surfaces in the room temperature fractured Nb:STO. Another possible explanation is that the fracturing behavior of Nb:STO at 50 K is different from that at room temperature due to different structural phases. Note that STO goes through structural phase transition at ~105 K from high temperature cubic phase ($Pm\overline{3}m$ (O_h^1)) to low temperature tetragonal phase ($I4/mcm$ (D_{4h}^{18})) [69]. Further temperature-dependent fracturing experiments will shed light on getting insights on the temperature effects on fracturing Nb:STO.

On the other hand, atom manipulations were reported for many different types of surfaces using STM tip and were also reported for the fractured Nb:STO surfaces [60]. The SrO terminated surface on the large terraces is the ideal playground for the manipulation by electric field applied at the STM tip-sample junction. With the set point of 1.4 V; 50 pA for imaging, the ability of the tip manipulation as a function of the electric pulse bias and duration is shown in Fig. 5.8a for topography and Fig. 5.8b for STS contrast [60]. By analyzing the topography and the dI/dV mapping, the depths and widths of the holes, threshold condition of creating holes, and the threshold of seeing observable dI/dV contrast change are summarized in Fig. 5.8c. Further testing revealed that the tip-sample electric field pulses could remove the SrO clusters away from the sample surfaces onto the tip or into the vacuum, leaving holes with half-integer unit cell height changes (~0.2 nm) [60]. The change of the STS contrast is originated from the exposed underlying TiO_2 terminated surface after the SrO clusters are removed [60]. It is interesting to note that the change of STS contrast, as shown in Fig. 5.8b, does not always follow the topography change. This was explained by the finite tip size effect and Smoluchowski smearing effects [60].

In addition to the hole creation (positive bias manipulation), the negative bias can re-deposit the SrO clusters from the tip back to the desired locations on the surfaces [60]. As shown in Fig. 5.8d–f, the holes can be refilled by the negative pulse procedure while the tip is positioned to the hole. The cluster deposition could be done anywhere on the surfaces as long as the tip was previously treated with the positive bias pulses (attracted SrO clusters on it). This further indicates that the tip can be conditioned by the negative bias pulses as the process to clean tip. This is very

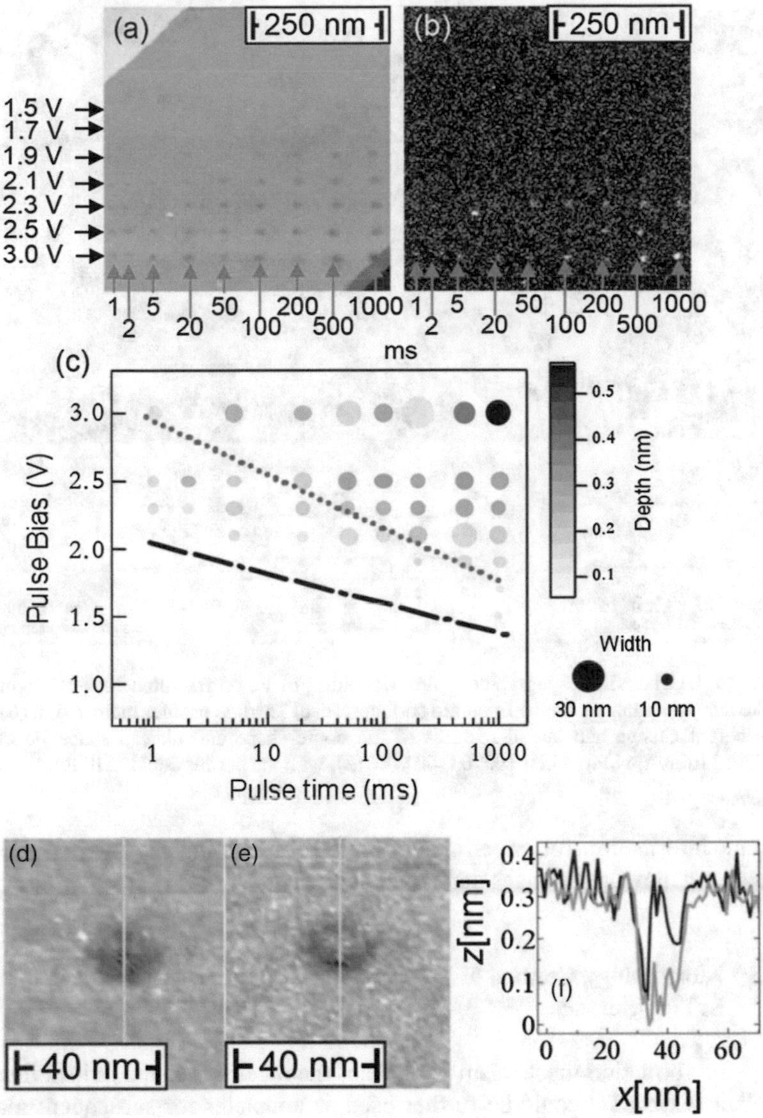

Fig. 5.8 (**a**) STM topography and (**b**) d*I*/d*V* mapping of STM-tip-modified fractured Nb:STO surfaces. Surface modifications were done with a systematically changed tip-sample bias and duration, as labeled in the images. (**c**) The results of the analysis of the dot features on the modified Nb:STO surfaces. (**d**, **e**) The STM topography of the hole and protrusion in the hole on the Nb:STO surfaces due to opposite polarities of the tip-sample bias. (**f**) The line profiles of the hole and the protrusion in the hole shown in (**d**, **e**). Reproduced from Appl Phys Lett 95, 163,107 (2009), with the permission of AIP Publishing

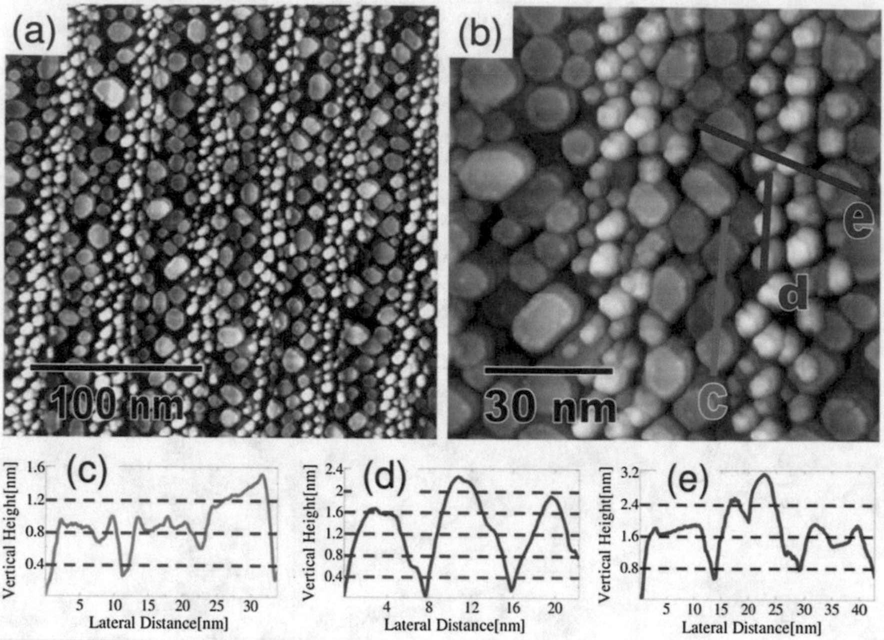

Fig. 5.9 (**a**, **b**) The STM topography of the deposition of Fe on fractured Nb:STO, which has strip-like dual terminations on the fractured surfaces. (**c**–**e**) The line profiles indicated in (**b**) showing the height change and lateral features of the dome shape and plateau shape Fe clusters. Reproduced from Appl Phys Lett 100, 031601 (2012), with the permission of AIP Publishing

important information for experiments aiming for probing the interfaces, which requires long moving distances that could degrade the tip.

5.3.1.3 Morphology Control of Fe Deposited on the Fractured Nb-Doped $SrTiO_3$ Surfaces

Though the formation mechanism is unclear, the dual terminated stripes found on the fractured Nb:STO could be further used as templates for subsequent material depositions. As demonstrated by Chien et al., the striped, dual-terminated Nb:STO surfaces were used to control the morphology of the Fe films deposited on it [64]. As-deposited Fe films exhibit nano-dots morphology with an average dot size/separation to be around 3.7 ± 1.0 nm. Upon annealing at 650 °C for 10 min., the Fe film morphology changed dramatically into two distinct types, as shown in Fig. 5.9. One is a dome shape with an average diameter of 8 nm; while the other is a larger (10–15 nm) plateau-like islands. It is believed that this change of the morphology upon annealing is driven by the interfacial energy difference between Fe/SrO and Fe/TiO_2 interfaces [46]. The change of morphology on the stripe-patterned substrate upon the annealing was ruled out with the annealing experiment before the Fe is

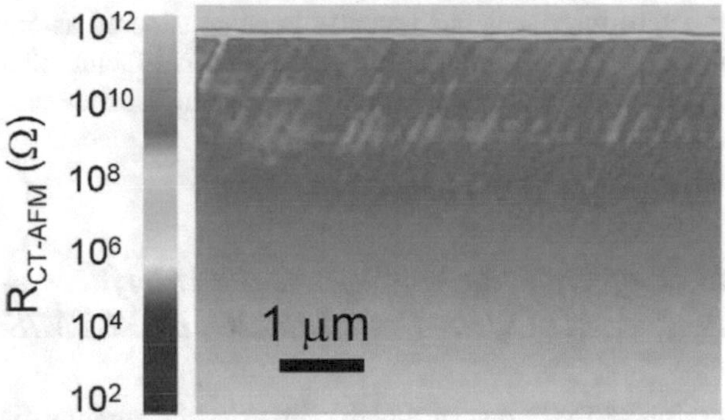

Fig. 5.10 XcAFM image of ion irradiated STO in cross-sectional geometry. A highly conductive region is visualized at ~50 nm depth after the ion irradiation. In addition, a decreased resistance when moving away from the surfaces is found to be in the range of few micrometers scale. This long range low resistance regions are considered as the evidence of the oxygen vacancy diffusion into the materials after the ion irradiation. Reproduced from J Appl Phys 107, 103704 (2010), with the permission of AIP Publishing

deposited. This finding points to the possibility of creating novel striped features of materials in the order of tens of nm scale. Further experiments on this directions may be fruitful of creating new material systems.

5.3.2 *Oxygen Vacancies and Their Migration of the Ion-Irradiated $SrTiO_3$ Surfaces*

Oxygen vacancy is one of the most important factors of controlling oxide properties. Probing the oxygen vacancy profile in depth is crucial for the further understanding of the physical phenomena at surfaces and interfaces. Ion irradiation or sputtering processes followed by post-annealing are common methods of cleaning solid material surfaces for various purposes. Naively, for oxides, one would assume the post-annealing process could heal the surface structures after the damages made by the ion-irradiations and is when the oxygen vacancies formed. However, Herranz et al. performed a cross-sectional conducting atomic force microscopy (XcAFM) measurement and revealed that the high concentration of the oxygen vacancies was induced ~50 nm near the surfaces after the Ar^+ irradiation on STO [70]. Furthermore, the oxygen vacancies could further diffuse into the materials up to micrometer scale [70]. In the Herranz's study, the oxygen vacancy-induced conductance was mapped with XcAFM. As shown in Fig. 5.10, at depth ~ 50 nm, lowest resistance was revealed, which is induced by the oxygen vacancies created by the ion irradiation. The resistance map also showed a gradual increase with length scale up to few micrometers, which is believed due to the oxygen diffusion length scale into the materials [70]. This finding pointed out that the preparation procedure for perovskite oxides should be carefully considered for subsequent purposes. The oxygen vacancy near the oxide hetero-interfaces has recently been considered as one of the major engineering topic for controlling complex oxide interfacial phenomena. This method could potentially be used to study the oxygen vacancy profile across the oxide hetero-interfaces.

5.3.3 Band Diagram Mapping Across the Interfaces of $La_{2/3}Ca_{1/3}MnO_3$/Nb-Doped $SrTiO_3$ by XSTM/S

The first successful XSTM/S measurement on perovskite complex oxide interfaces was achieved by Chien et al. on $La_{2/3}Ca_{1/3}MnO_3$/Nb-doped $SrTiO_3$ (LCMO/Nb:STO) system [23]. The sample preparation used in this study is the controlled fracturing process describe earlier. One of the key issues for getting nice interface regions after fracturing is the fracturing direction. Since the fracturing is done by introducing a bending strain by moving the scribed samples against a rigid cleaver, the fulcrum side of the sample is subject to be severely damaged. Thus, the film/interface side of the sample cannot be placed as the fulcrum side. On the other hand, if the film/interface is placed on the initial cracking side, since there is no initial weak notch, the fracturing results are also not satisfied. The best way of doing the fracture would then be dicing the samples from the side, which damaged some part of the film, but leaves the rest of the film intake. Then fracture the samples from sideways. The geometry of the fracturing, diced side, and the sample orientations are shown in Fig. 5.4. After the fracture is successful, the next challenge is to find the interfaces with the STM tip. This problem can basically be resolved by having a scanning electron microscopy (SEM) in the STM system, so that the tip position could be easily manipulated to land on the region near the interfaces. However, since the typical STM system does not have an SEM with it, the STM tip can only land on the fractured surfaces using optical microscope equipped in STM system with knowing hundreds of micron meters away from the interfaces. When "walking" toward the interfaces, the tip moving speed should be kept slow while having the constant current feedback loop on since the fractured surfaces may have significant height change beyond the range of the z-piezo movement capability. In the case of Nb:STO, parameters of the set point as 3.0 V and 50 pA are found to be working well while the cruising speed is set to be 100 nm/s. The tip-sample distance is subject to be adjusted in micron scale during the long walk.

Figure 5.11 shows the STM, STS, and SEM images of the fractured LCMO/Nb:STO interfaces. In Nb:STO region, the dual-terminated surfaces extended all the way to the interface, as shown in Fig. 5.11b, where the dI/dV contrast can be used to distinguish SrO and TiO_2 terminated regions [59]. In LCMO region, the surface does not show clear terraces with steps. This is likely related to the strained LCMO and/or the strain relaxation during the fracturing [23]. The STS contrast at 3.0 V could be explained with the dI/dV point spectra on LCMO, SrO, and TiO_2 terminated Nb:STO surfaces, as shown in Fig. 5.11d. At positive bias, the spectral weight

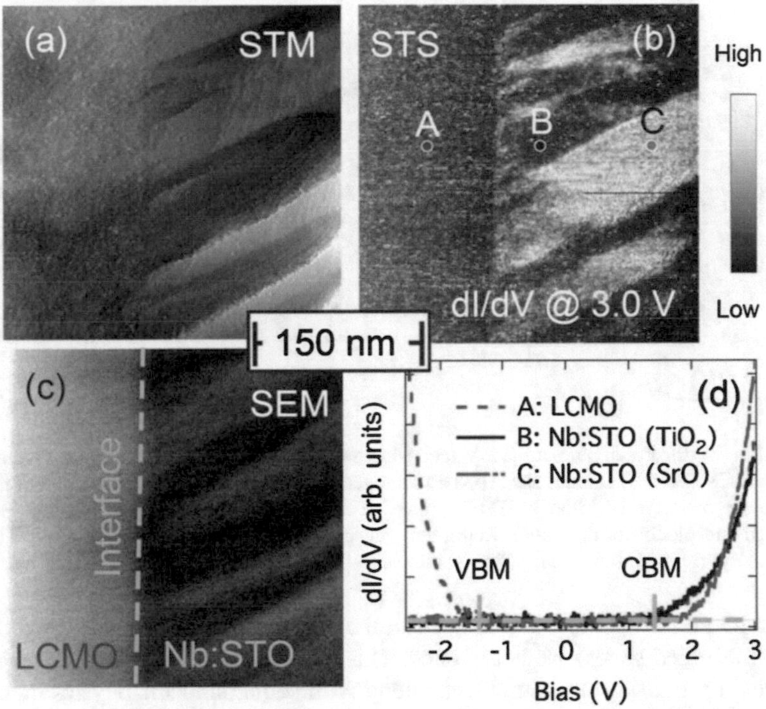

Fig. 5.11 The interfacial region of LCMO/Nb:STO measured with (**a**) STM topography; (**b**) dI/dV mapping; and (**c**) SEM. The locations of the interfaces are clearly visible. (**d**) The dI/dV point spectra measured on LCMO, and two different terminations on Nb:STO were plotted. LCMO exhibited a clear smaller energy band gap compared to Nb:STO materials. Reprinted with permission from Phys Rev B 82, 041101(R) (2010). Copyright 2010 American Physical Society

is similar for three types of surfaces; while the SrO terminated surface has slightly higher value at 3.0 V, which makes it to be brightest in Fig. 5.11b. It is also clear that in the negative bias region, the spectral weight of LCMO is much higher than that of STO (both SrO and TiO_2 terminated surfaces) due to the smaller energy gap in LCMO. Thus, it would be clear to distinguish LCMO from STO in STS using negative bias. However, since the DOS of STO is very low in negative bias, the scanning is not stable while maintaining constant current. Instead, Chien et al. measured 30 × 30 grid dI/dV spectra in a region near the interface with the scale of 300 nm × 300 nm. Using this grid spectrum, the spatial dI/dV contrast at −2.5 V could be reproduced, as shown in Fig. 5.12a. The locations of the interfaces then can be easily identified. Furthermore, the grid dI/dV spectra could be averaged into different spectra as a function of the distance to the interfaces. By analyzing the data, the conduction band minimum (CBM) and valence band maximum (VBM) at each location could be extracted. In other words, the electronic band diagram across the interface could be mapped directly from the experimental XSTM/S data, as shown in Fig. 5.12b. One of the striking results from this XSTM/S study is that the bands

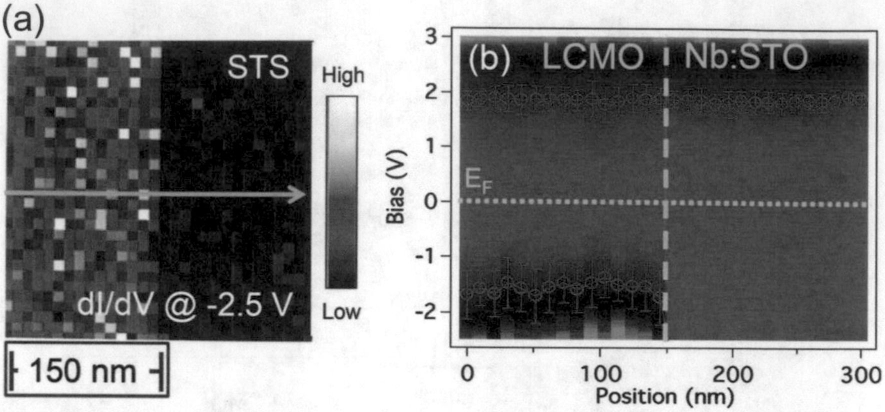

Fig. 5.12 (**a**) d*I*/d*V* mapping at −2.5 V reconstructed from d*I*/d*V* point spectra taken in an array near the LCMO/Nb:STO interfaces. (**b**) The averaged d*I*/d*V* point spectra as a function of the distance away from the LCMO/Nb:STO interfaces. The conduction and valence band edges were extracted and plotted in the figure. Reprinted with permission from Phys Rev B 82, 041101(R) (2010). Copyright 2010 American Physical Society

in LCMO and in Nb:STO are aligned in the CBM (unoccupied states), while the VBM (occupied states) are not aligned. This band alignment at the CBM was not expected by transport measurements along with semiconductor physics modeling [71]. This indicates that the re-examination of the validation of using semiconductor physics, which is based on weak correlation and coupling assumption, on strongly correlated/coupled complex oxide materials is needed. Note that there is no observable band bending near the interfaces longer than 10 nm, which is the spatial resolution in the data [41]. Nevertheless, one important message from this study is that the XSTM/S could be utilized for studying the band diagram across the interfaces of complex oxides.

5.3.4 *Highly Mobile Two-Dimensional Electron Gas at Interfaces of $LaAlO_3/SrTiO_3$ by Cross-Sectional Conducting AFM and STM*

In the recent decade, two-dimensional (2D) electron gas (2DEG) has been observed at $LaAlO_3/SrTiO_3$ (LAO/STO) interfaces [28, 40, 41, 53, 54, 72, 73], which opened a research field aiming at complex oxide interface engineering and physics. Two interfacial configurations are found to be crucial to induce this 2DEG: (a) STO termination and (b) LAO thickness. In the former, the interfacial configurations have to be $[LaO]^{+}[AlO_2]^{-}$ planes on $[SrO]^{0}[TiO_2]^{0}$ or $[LaO_3]^{3-}[Al]^{3+}$ planes on $[SrO_3]^{4-}[Ti]^{4+}$ for (001) or (111) orientation, respectively, to induce the 2DEG [74]. For the later, the grown LAO film needs to at least exceed certain critical thickness,

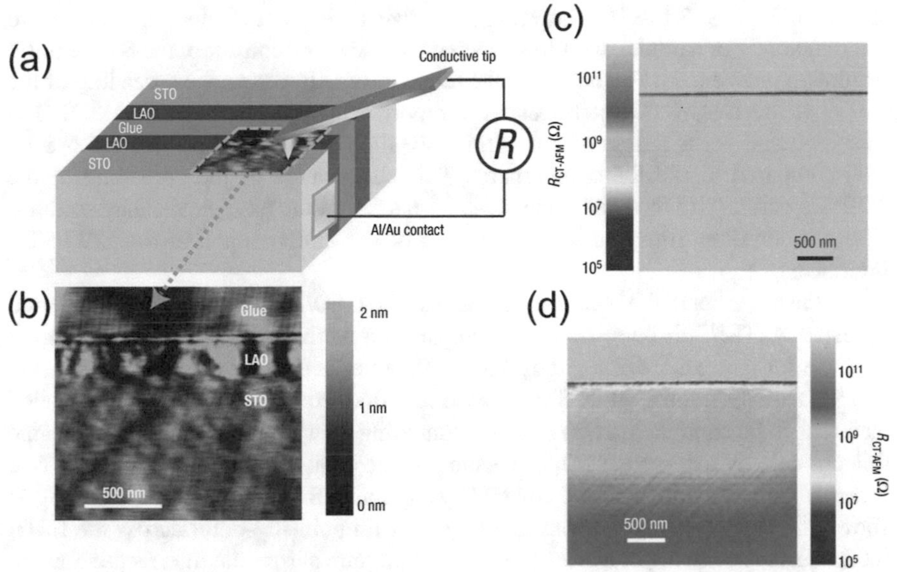

Fig. 5.13 (**a**) Schematic of XcAFM measurements for LAO/STO interfaces. Two LAO/STO samples were glued face-to-face followed by polishing prior to XcAFM measurements. (**b**) XAFM topography showing the locations of the LAO, STO, and glue. (**c**) Resistance mapping measured simultaneously with the AFM topography. A highly conductive region with thickness of ~7 nm was revealed at the LAO/STO interfaces. Oxygen vacancies were eliminated during the sample growth. (**d**) Resistance mapping in cross-sectional geometry with LAO thin film grown on oxygen-deficient STO substrate. Similar highly conductive interfaces are revealed. In addition, a micrometer scale lower resistance regions are observed due to the oxygen vacancies. Reprinted with permission from Nat Mater 7, 621 (2008). Copyright 2008 Nature Publishing

typically ranging from 3 to 8 unit cells depending on the crystal orientation, to have this 2DEG to be created [75]. The mechanism of the formation of the 2DEG at the LAO/STO interfaces is still controversial. Three major and most popular explanations are: (1) polar catastrophe [15]; (2) oxygen vacancies [72]; and cation interdiffusion [76, 77]. One of the first direct evidences of the existence of the 2DEG at LAO/STO interfaces was the use of the cross-sectional conducting atomic force microscopy (XcAFM) measurements [53, 54], where high conductive interfaces were revealed. The successful XcAFM measurements were achieved by preparing the LAO/STO samples through gluing two samples face-to-face together and then polishing in cross-sectional geometry [53], as shown in Fig. 5.13a, b. The resistance mapping in cross-sectional geometry on LAO/STO system revealed two important information for clarifying the 2DEG mechanism. First, as shown in Fig. 5.13c [53], the metallic electron gas was first confirmed at the LAO/STO interfaces within the length scale of ~7 nm at room temperature. In this sample, the oxygen vacancy was eliminated by in-situ annealing in an oxygen-rich environment [41]. Based on this controlled oxygen vacancy elimination sample, the oxygen vacancy-induced conductance was excluded. On the other hand, in the oxygen-deficient STO samples, as shown in Fig. 5.13d, STO showed conductive nature over a length scale up to micrometers. For samples with low oxygen vacancy concentration, the STO exhibit insulating nature [53]. Furthermore, the temperature effects on the spreading of the 2DEG at the LAO/STO interfaces were reported. The spreading of the 2DEG is increased and yet is still confined within ~10 nm scale at low temperature (~8 K) [54] compared to the RT counterparts. This short range of the spreading of the 2DEG is contrary to the prediction of ~50 nm scale in the oxygen vacancy scenario [72], further supporting the intrinsic nature of the 2DEG found at the LAO/STO interfaces.

On the other hand, the band diagram across the LAO/STO interfaces was reported by using XSTM/S through fracturing preparation methods [26]. The most difficult obstacle for using STM/S to study LAO/STO interfaces is the fact that both LAO and STO are insulators; while STM/S requires conductive materials for the tunneled electrons to be conducted away from the tunneling locations. Huang et al. overcame this obstacle by using Nb:STO as the substrate to grow high-quality STO thin film followed by LAO thin film, and covered by a metallic $SrRuO_3$ capping layer [26], as shown in Fig. 5.14a. By measuring STS spectrum point-by-point across the LAO/STO interfaces, Huang et al. revealed a band diagram across the interfaces showing the existence of electric field inside 5-unit-cell-thick LAO thin films, as can be seen in Fig. 5.14b, c [26]. What would be interesting to see is to make the similar measurements while applying external electric bias across the LAO/STO interfaces with the LAO thickness slightly above or below the critical thickness (3 or 4 unit cells) to observe how the internal electric field in LAO layer evolves accordingly. This type of measurements could provide greater insights into how to modify and control the observed 2D electron gas at LAO/STO interfaces [78].

5.3.5 Domain Walls in $BiFeO_3$ and Interfaces of $BiFeO_3$/Nb-Doped $SrTiO_3$ Studied by XSTM/S

One of the most important functionalities in complex oxide materials is the multiferroics, in which more than one ferroic properties are presented in the systems. For example, $BiFeO_3$ (BFO) has ferroelectric and ferromagnetic properties coexist in one phase [8, 79–81]. The ferroelectric and ferromagnetic type of multiferroics (magnetoelectric multiferroics) suggests possible applications, such as transducers, attenuators, filters, field probes, and data recording devices based on electric control of magnetization and vice versa [82]. BFO is one of the most studied single-phase magnetoelectric multiferroic materials. BFO has electric polarization directing along diagonal of the pseudo-cubic/rhombohedral axis; while the easy axis of the magnetic polarization is perpendicular to the electric polarization, as shown in Fig. 5.15 [79]. Based on the symmetry of the crystal structure, these polarization directions have same local minimum energy. In other words, in BFO materials, the eight polarization directions (positive and negative of four diagonal directions) are

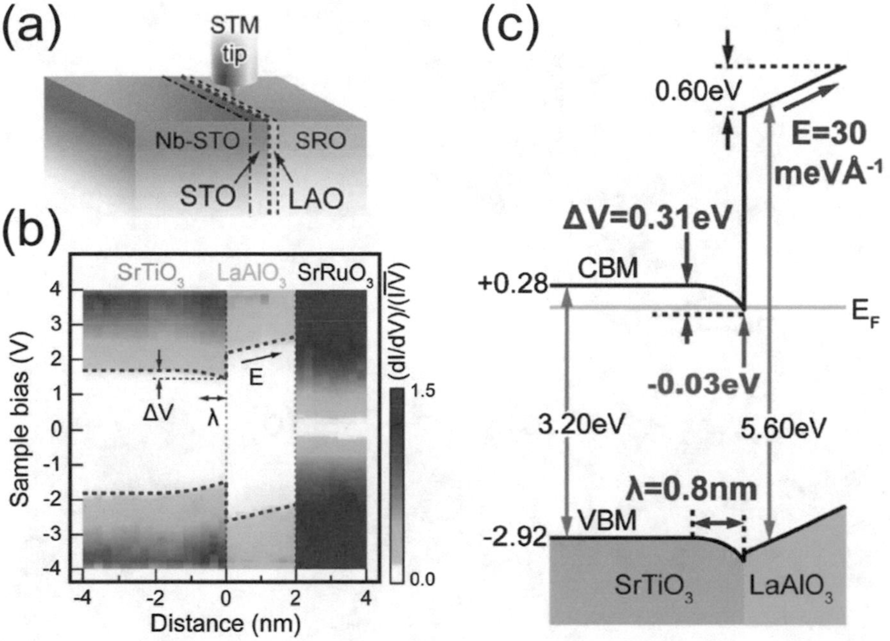

Fig. 5.14 (**a**) Schematic of XSTM/S measurements on the LAO/STO system. In particular, Nb:STO and SRO were used as conductive substrate and capping layer, respectively, to study the interfaces of the insulating LAO and STO materials. (**b**) d*I*/d*V* point spectra taken across the interfaces were observed by the XSTM/S measurements. (**c**) Electronic band diagram across the LAO/STO interfaces are extracted from experimental data. Reprinted with permission from Phys Rev Lett 109, 246,807 (2012). Copyright 2012 American Physical Society

energetically degenerate. The possible angles between each pair of the polarization directions are 71°, 109°, and 180° [79]. Similar to $PbZr_{1-x}Ti_xO_3$ (PZT) films, when BFO thin films are grown on cubic substrate, such as STO, domain patterns may develop with either {100} (109° domain wall) or {101} (71° domain wall) boundaries upon cooling from growth temperature (above Curie temperature, when the materials are in cubic paraelectric phase) to RT (below Curie temperature, when the materials are in rhombohedral ferroelectric phase) [83]. Possible domain patterns in the thin films on STO can be expected [52]. This orientation of the domain patterns is visible in the cross-sectional view of the thin films.

Domain walls in ferroic materials are considered as two-dimensional (2D) topological defects [52], which play an important role in determining the functionality of the crystals [84]. For example, the ferroelectric polarization and magnetization could be controlled by both magnetic and electric fields, respectively, in $GdFeO_3$ [85]. In BFO, the domain walls in BFO are the sources of the exchange bias interaction between the BFO and the adjacent ferromagnetic metal layer [86]. The 109° domain walls were reported to be the contribution for uncompensated spins [87] and were reported to be able to enhance the electrical conductivity in BFO [88, 89]. To

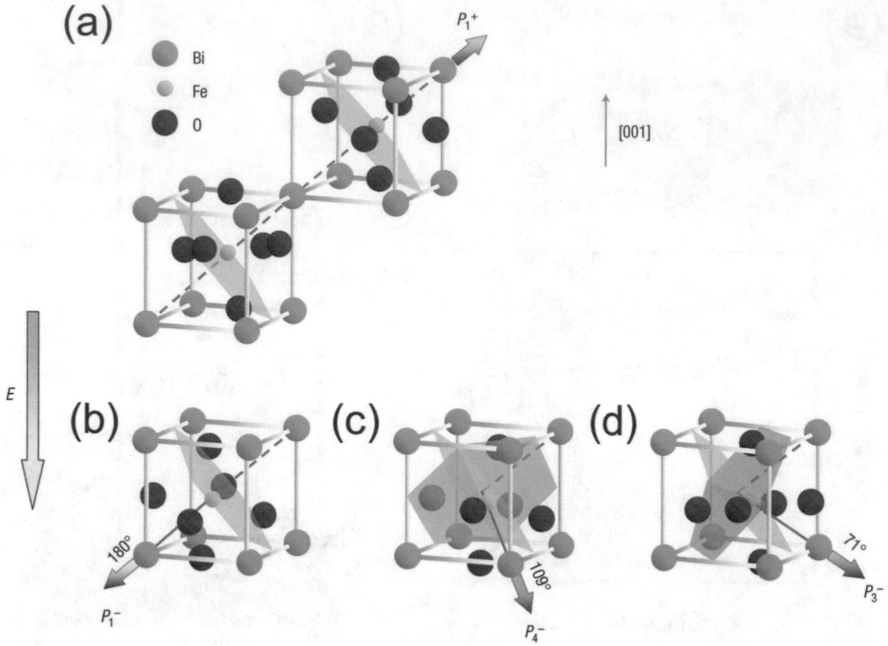

Fig. 5.15 Schematics of the crystal structures of BFO and their electric polarization orientations. (**a**) The crystal structure and one of the ferroelectric orientation. (**b–d**) The relationships between the other ferroelectric orientations with the first one shown in (**a**). Three distinct angles between the ferroelectric orientations are 71°, 109°, and 180°. Reprinted with permission from Nat Mater 5, 823 (2006). Copyright 2006 Nature Publishing

gain further insights on the origin of the high conductivity at the domain walls in BFO, Chiu et al. utilized STM/S to study the domain walls in as-grown BFO from cross-sectional geometry [52]. The types of domain walls—71° DW and 109° DW—could be easily distinguished from each other in STS contrast mapping based on the orientation of the domain walls in cross-sectional geometry. By measuring *dI/dV* point spectra along a line crossing the domain walls, the band edges and the band gaps are extracted experimentally for both types of domain walls [52]. It is revealed that the 109° domain walls exhibited observable lower energy gaps at the domain walls, which have a length scale of ~2–3 nm. It is argued and believed that the lower band gaps found at the 109° domain walls are the origin of the high conductivity associated with this type of domain walls.

Since BFO has ferroelectric polarization properties and the XSTM/S has the ability to map out the electronic band diagram across an interface, Huang et al. demonstrated that the band bending at the BFO/Nb:STO interfaces could be controlled by the BFO polarization configurations [27]. As shown in Fig. 5.16 [27], two BFO/Nb:STO samples with down polarized (P-down) and up polarized (P-up) are prepared for XSTM/S measurements. The as grown BFO/Nb:STO thin film has a P-down polarization; while the P-up samples were prepared by applying an up-direction static electric field on the as grown BFO/Nb:STO thin films. The band

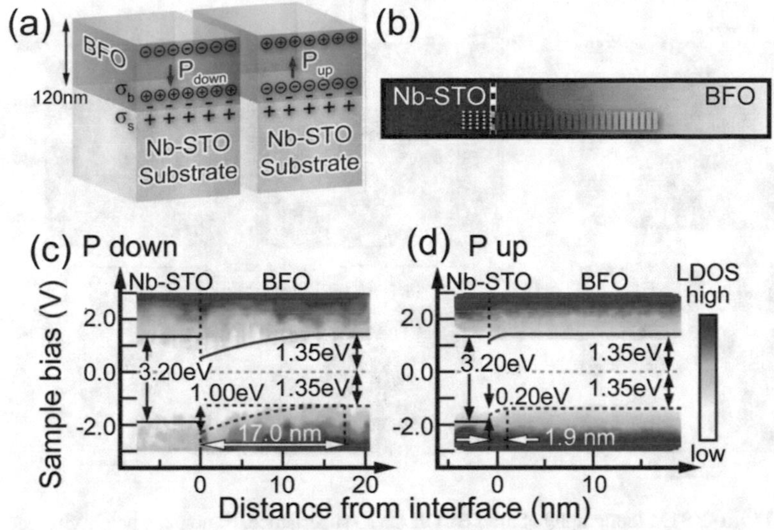

Fig. 5.16 (**a**) Schematic showing how the polarity of the ferroelectric polarization in BFO thin film may affect the charge distribution near the interfaces in contact with Nb:STO. (**b**) Points where d*I*/d*V* spectra were measured across the BFO/Nb:STO interfaces. (**c**, **d**) The d*I*/d*V* spectra taken as function of the distance away from the BFO/Nb:STO interfaces with the ferroelectric polarization orientation pointing down and up, respectively. The corresponding electron band bending was revealed through the visualization of the electronic band diagram across the BFO/Nb:STO interfaces. Reproduced from Appl Phys Lett 100, 122903 (2012), with the permission of AIP Publishing

diagrams across the BFO/Nb:STO interfaces of these two samples were mapped out by the XSTM/S. The altered electronic band bending near the BFO/Nb:STO interfaces were clearly observed, as shown in Fig. 5.16c, d. The widths of the depletion regions were found to be changed from 17.0 nm for P-down to 1.9 nm for P-up samples; while the built-in potentials were found to be changed from 1.0 V for P-down to 0.2 V for P-up samples. The key message in this study is that the XSTM/S is an ideal tool not just to study the band diagram across the interfaces of complex oxide, but it could be actively used as the tool to study the electronic structures upon the external stimuli. This opens the door for using XSTM/S to wider applications and material systems.

5.3.6 Short Range Charge Transfer at Interfaces of $YBa_2Cu_3O_{7-\delta}/La_{2/3}Ca_{1/3}MnO_3$ Superlattice by XSTM/S

The interactions between ferromagnetic materials (FM) in contact with the superconducting materials (SC) have invoked a long history of research [90]. The main focus was on the interplay between the ferromagnetic order and the formation of the singlet cooper pairs in superconductors. The former favor the parallel spins while

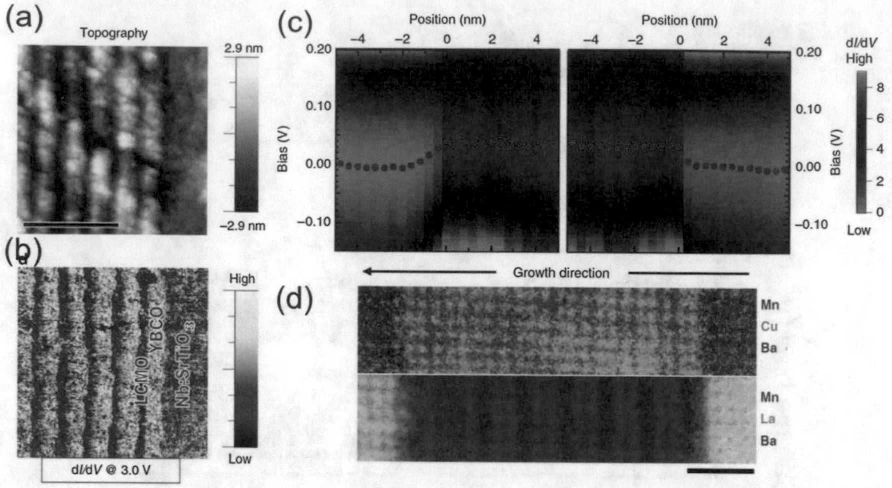

Fig. 5.17 (**a**) XSTM topography of the YBCO/LCMO superlattice. Nanometer height changes across different materials were observed. (**b**) The d*I*/d*V* mapping recorded simultaneously with the topography measured in (**a**). YBCO and LCMO could be unambiguously distinguished with the d*I*/d*V* contrast. (**c**) d*I*/d*V* point-by-point spectra measured across interfaces with different growing sequence—YBCO on LCMO or LCMO on YBCO. The electronic structures exhibited different spatial transition length scales, however all are under 1 nm in both cases. (**d**) The STEM with EELS data revealed that the different transition length scales found in d*I*/d*V* spectra were originated from the different atomic intermixing depending on the growing sequence. Reprinted with permission from Nat Commun 4, 2336 (2013). Copyright 2013 Nature Publishing

the later favors the anti-parallel spins. The competing orders near the interfaces of the SC and FM result in lowering the superconducting transition temperature due to the diffusion of the FM ordering into the SC; while, on the other hand, the superconducting current may propagate through the FM with certain thickness [91, 92]. The understanding of the FM/SC interplay between elemental FM and conventional SC was well understood [90]; while that in oxide-based FM/SC is still controversial on many topics. One of the heavily debated topics is the length scale of the charge transfer across the interfaces [16, 24, 93].

The charge transfer length scale at the $YBa_2Cu_3O_{7-\delta}/La_{2/3}Ca_{1/3}MnO_3$ (YBOC/LCMO) interfaces was studied and directly revealed by Chien et al. using XSTM/S [24]. This is achieved by probing the built-in potential, a resulting consequence of the charge transfer between two dissimilar materials across the interfaces. As shown in Fig. 5.17a, b, Chien et al. successfully probed the YBCO/LCMO superlattice in cross-sectional geometry [24]. By measuring the point-by-point d*I*/d*V* spectra across many interfaces shown in Fig. 5.17a, b, the band diagrams across the interfaces were revealed. Furthermore, the charge transfer length scale across the YBCO/LCMO interfaces were visualized by analyzing the spatial evolution of the d*I*/d*V* spectra, as shown in Fig. 5.17c. Interestingly, depending on the growth sequence, the charge transfer length scale across the YBCO/LCMO interfaces is found to be ~0.26 nm and ~0.96 nm for YBCO on LCMO and for LCMO on YBCO, respectively. This difference in charge transfer length scales in different growth sequences was further explained by the different levels of ionic intermixing across the interfaces which were confirmed by scanning transmission electron microscopy (STEM) and electron energy loss spectroscopy (EELS) data, as shown in Fig. 5.17d [24]. Note that the interface roughness depends on the sequence of growing which has been reported in many other hetero-oxide systems [94–96]. These results indicate that the length scale of the charge transfer between YBCO and LCMO is in the order of 1 nm or less [24], which is considered as the short range charge transfer scenario. This study demonstrates that the XSTM/S technique could be used to directly resolve the charge transfer length scale across the interfaces of dissimilar materials, and the results of this study put an upper limit of the charge transfer length scale for YBCO/LCMO material systems to be ~1 nm, which could be further used to understand the proximity effects in the oxide-based FM/SC interfaces.

5.3.7 *Schottky Built-in Electric Field Altered Fracture Toughness in Nb:STO at Interfaces in Contact with $LaNiO_3$*

Electric field effects on mechanical properties are rarely explored in solid materials. Based on the different conductivity nature of the materials, varieties of mechanisms were proposed to explain the electric field altered mechanical property. For example, conductive materials will screen electric field to result in zero field inside them, thus (1) change of the valence electron density [97], and (2) the escape of vacancies from the grain interior to the grain boundary [98, 99], were proposed to explain the electric field effects on the mechanical properties in conductive materials. However, there is no overall consensus among different conductive materials in different reports [97–103]. For polar materials, such as ZnS, (3) the interactions between the electric field and the charged dislocations are responsible for the decreases of the hardness [104]. On the other hand, in non-polar materials, such as SiO_2, (4) the change of the inter-atomic bond lengths due to the dielectric response ($\vec{P} = \varepsilon_0 \chi_e \vec{E}$) results in the increase of the hardness [105]. And finally, for piezoelectric materials, such as lead zirconate titanate (PZT), (5) the domain switching due to the electric field results in decreasing the fracture toughness [106]. All of above reports were focused on bulk materials. There is no discussion for the mechanical properties of nano-materials upon the applications of the electric field. This type of topic is very important for nanoelectromechanical system (NEMS) devices, in which strong electric fields are typically utilized in controlling the nano-materials. At complex oxide interfaces, the strong built-in electric field may also play an important role in altering the mechanical properties near the interfaces. To reveal this effect, Chien et al. chose the $LaNiO_3$/Nb:STO system to perform the XSTM/S, in which a fracturing process was inherent in the sample preparation [62].

STO is neither a polar nor a piezoelectric material, but it has a strong electric field and temperature-dependent permittivity [107]. This property implies that the polarization density, $\vec{P}$, has a nonlinear function with respect to the electric field. Since the polarization density is directly related to the atomic displacement in a unit cell, the inter-atomic bond length is expected to be directly affected by the application of the electric field, hence the change in mechanical properties, such as fracture toughness. On the other hand, $LaNiO_3$ (LNO) is a metallic oxide material at all temperature tested [29]. Thus, a Schottky barrier with the strong built-in electric field at the LNO/Nb:STO interfaces is expected.

This subtle change of the mechanical properties has been observed in Nb:STO in contact with metallic LNO using XSTM/S [62]. Following the reported controlled fracturing procedure of XSTM/S for complex oxides, a trench-like morphology was revealed at the interfaces of Nb:STO in contact with LNO, as shown in Fig. 5.18a. The trench has a depth of ~0.6 nm, which equals to 1.5 unit cell height, and a width of ~6 nm, which was found to be closely related to the depletion width formed at the Nb:STO/LNO interfaces. The built-in potential was also visible in the d*I*/d*V* mapping measured simultaneously with the topography, as shown in Fig. 5.18b. The d*I*/d*V* contrast measured at 3.0 V is highest in the trench and gradually decay into the Nb:STO substrate. The subtle changes in the d*I*/d*V* signals in Nb:STO as a function of the distance to the interfaces could be easily observed in Fig. 5.18c. The band diagram, hence the Schottky barrier profile, was revealed by the point-by-point d*I*/d*V* spectra taken across the LNO/Nb:STO interfaces, as shown and depicted in Fig. 5.18d. With in-depth analysis, Chien et al. concluded that the spatial evolution of the electric field in the depletion region is the origin of the alteration of the atomic bond length, hence the mechanical property—fracture toughness—of the Nb:STO near the interfaces. The same mechanism of the non-linear response of the polarization density is also argued to be responsible for the observed dielectric dead layer reported at the interfaces of STO in contact with various types of the metallic film [108].

5.3.8 *Interfaces of Y-Doped $BaZrO_3$/$NdGaO_3$ by Cross-Sectional Electrochemical Strain Microscopy (XESM)*

Ion migration phenomenon is one of the many interesting and useful phenomena of perovskite oxides related to the applications in the fuel cell field. Among the perovskite materials, Y-doped $BaZrO_{3-\delta}$ (BZY) has shown excellent chemical stability [109, 110] as well as high protonic conductivity (higher than oxygen ion conductors) in the temperature range of 300–600 °C [111], which are promising for the applications in protonic fuel cells. Later, very high values of conductivities in BZY thin film (tens of nm) grown on (110) $NdGaO_3$ (NGO) have been reported in the temperature range of 550–600 °C [112]. The very high lattice mismatch (10%) between the film and the substrate suggests that the heavily strained interfaces may

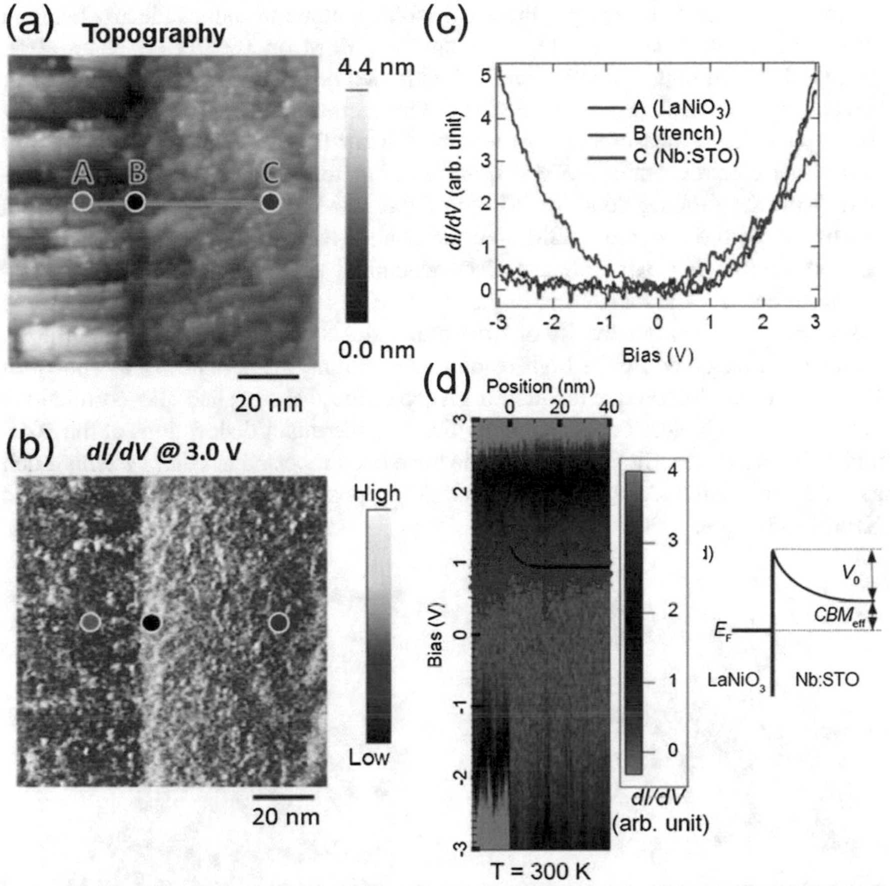

Fig. 5.18 (**a**) STM topography of LNO/Nb:STO in cross-sectional geometry. A topographic trench is clearly observable with width of ~6 nm and depth ~ 0.6 nm. (**b**) d*I*/d*V* mapping recorded simultaneously with the topography. The trench exhibited highest contrast and gradually decays into the substrate. (**c**) d*I*/d*V* point spectra measured at three locations indicated in (**a**). The similarity between the spectrum B and C infers that the trench is the Nb:STO. (**d**) Schottky barrier is visualized by the d*I*/d*V* point spectra measured point-by-point across the LNO/Nb:STO interfaces. Reprinted with permission from Sci Rep 6, 19017 (2016). Copyright 2016 Nature Publishing

be a key parameter to tailor defect densities in thin epitaxial films. To address this issue, Yang et al. utilized the electrochemical strain microscopy (ESM) to study the BZY/NGO interfaces in both in-plane and cross-sectional geometry [113].

The BZY/NGO samples were fractured prior to the XESM measurements (Fig. 5.19a) and the height variation of the fractured surfaces was determined by AFM to be in the order of ~10 nm. The ESM technique utilizes an SPM tip as an electrode in electrochemical reaction. When a sufficiently high tip-sample bias is applied, the electrochemical reaction is activated, in which the mobile ionic species between the tip and sample junction could be generated or annihilated. The electrochemical reaction thus changes the local molar volume to induce electrochemical strain in 2–5 pm level [114, 115]. Using the XESM on BZY/NGO, Yang et al. clearly observed higher electrochemical activities near the interfaces compared to the film regions, as shown in Fig. 5.19b, c. This same picture has been confirmed by the in-plane ESM measurements where the thinner BZY (20 nm thick) film exhibited higher electrochemical ESM responses compared to the 300 nm thick counterpart. With the probing depth of 20 nm of the ESM technique, this result is in a consistent picture with the XESM measurements—the interfaces of BZY and NGO are indeed showing extraordinary electrochemical activities. These high electrochemical activities associated with the BZY/NGO interfaces was confirmed to be associated with the high density of structural dislocations found near the interfaces, which were observed by the high-resolution scanning TEM (STEM), as shown in Fig. 5.19d. In this study, the authors successfully established the correlations between the high electrochemical activities and the misfit dislocations at the BZY and NGO interfaces. Similar phenomena have been reported in other systems, such as STO/MgO interfaces [116], which could be possibly further confirmed by the XESM technique.

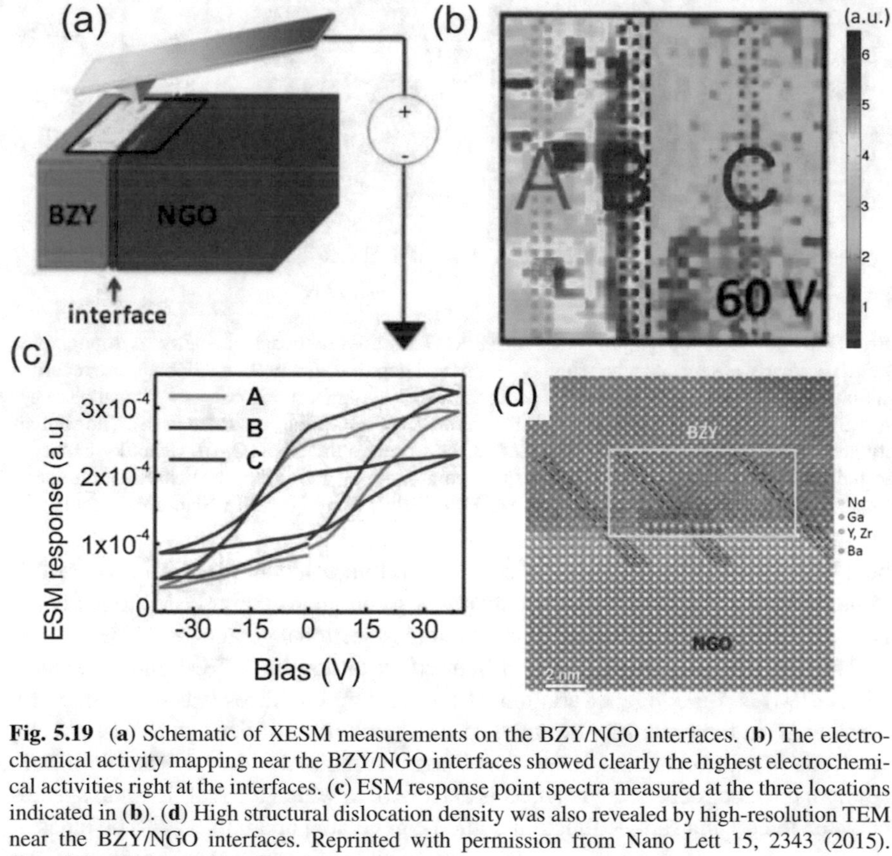

Fig. 5.19 (**a**) Schematic of XESM measurements on the BZY/NGO interfaces. (**b**) The electrochemical activity mapping near the BZY/NGO interfaces showed clearly the highest electrochemical activities right at the interfaces. (**c**) ESM response point spectra measured at the three locations indicated in (**b**). (**d**) High structural dislocation density was also revealed by high-resolution TEM near the BZY/NGO interfaces. Reprinted with permission from Nano Lett 15, 2343 (2015). Copyright 2015 American Chemical Society

5.3.9 Interfaces of $La_{0.65}Sr_{0.35}MnO_3/SrTi_{0.2}Fe_{0.8}O_3$ Created by Focused Ion Beam Milling

Contrast to the methods by fracturing and polishing, Kuru et al. utilized focused ion beam (FIB) milling to expose the interfaces of $La_{0.65}Sr_{0.35}MnO_3/SrTi_{0.2}Fe_{0.8}O_3$ (LSM/STF) superlattices for STM measurements [55]. The choice of LSM and STF was based on their potential applications in fuel cell field and their distinct electronic properties [117, 118], which could lead to discernable STM measurements. The FIB is tuned to hit the sample surface with a shallow angle to create an inclined surface with the interfaces in the superlattice exposed for STM measurements without going to cross-sectional geometry. The exposed interfaces were probed by both AFM and STM in which the two composited layers exhibit distinct morphology. In AFM, LSM layers have a roughness of ~2 nm; while the STF layers have less than 1 nm roughness. This was attributed to the different sputtering efficiency on different materials upon the FIB milling. More interestingly when comparing STM topography to the AFM topography, the STF layers exhibit dips with the depths to be in the order of ~10 nm in STM. Since AFM is sensing the atomic force directly, while the STM has an electronic origin, it is most likely that the AFM topography represents the true morphology, while the STM topography revealed the contrast due to the different electronic properties. This hypothesis was further explained by the measured *I–V* spectra in STF and LSM regions. At 0.5 V, LSM shows much larger tunneling current than STF does. When moving from LSM into STF regions, the tip has to move closer to the STF surfaces to maintain constant current feedback, thus dip appears in STF regions in STM topography. The authors argued that the surfaces obtained by the FIB do not show significant damage. This work demonstrated an alternative way to prepare interfaces for STM measurements without going into cross-sectional geometry. This method greatly eases the needs of skills of measuring samples in cross-sectional geometry and the possibility of degrading the tip after the long moving from the land point to the interfaces (typically few to tens of micrometers distance) in cross-sectional measurements.

5.3.10 Interfaces of $La_{0.8}Sr_{0.2}CoO_3/(La_{0.5}Sr_{0.5})_2CoO_4$ Created by Focused Ion Beam Milling

The FIB preparation method for interfacial STM studies was also used in $La_{0.8}Sr_{0.2}CoO_3/(La_{0.5}Sr_{0.5})_2CoO_4$ (LSC_{113}/LSC_{214}) multilayers [63]. The interests of this material system originated from the faster oxygen reduction reaction (ORR) kinetics in the LSC_{113}/LSC_{214} hetero-interfaces compared to single phases of either LSC_{113} or LSC_{214} at 500 °C [119, 120]. It is very technically important to search for materials or material systems with high ORR activity below 700 °C for solid oxide fuel cells (SOFCs) applications. LSC_{113} has been studied intensively as the cathode material for the applications of SOFCs due to its high electric and ionic conductivity

[121, 122]. The ORR activity on LSC_{113} is found to be limited by the availability and the mobility of the surface oxygen vacancies [123]. LSC_{214}, however, exhibits undesired low electrical conductivity but desired high oxygen diffusion kinetics [124]. Though the high ORR activities were reported in the LSC_{113}/LSC_{214} hetero-interfaces, the microscopic understanding is still limited.

To gain insights into the mechanism of this LSC_{113}/LSC_{214} multilayers exhibiting extraordinary ORR activities, Chen et al. utilized FIB to prepare samples for STM studies [63]. The two structural phases were clearly distinguished from each other based on the tunneling spectroscopy measurements. Both LSC_{113} and LSC_{214} showed energy gaps at room temperature. In particular, LSC_{113} has a smaller gap, consistent with the higher conductivity, compared to the LSC_{214} counterpart. Interestingly, the tunneling spectra measured at elevated temperatures (250–300 °C) exhibited zero band gaps (metallic phase) in both LSC_{113} and LSC_{214} regions. For comparison, in the single-layer LSC_{214} thin film, finite energy gaps were measured at the same elevated temperatures, indicating the zero band gaps found in multilayer is a result of the influence of the neighboring LSC_{113}. Chen et al. explained it as the result of the interactions of the oxygen vacancies across the LSC_{113}/LSC_{214} interfaces. This result indicates that the electronically activated (toward metallic behavior) LSC_{214} at intermediate temperature may have overcome the disadvantage of the low electronic conductivity in the single-phase LSC_{214} materials. The vicinity to the LSC_{113} serves an important factor to activate LSC_{214} electronically. This observation and the proposed mechanism provide a great insight on how the LSC_{113}/LSC_{214} interfaces may have extraordinary ORR activities and how STM can provide insightful information in the SOFC field.

5.4 Summary and Perspectives

This article reviews recent developments of the XSPM measurements on complex oxide interfaces with a focus on the research topics that could be studied with XSPM in an unprecedented insight. Among the reported studies, polishing, FIB milling, and fracturing are the three major methods successfully preparing the interfacial regions for XSPM measurements. Various types of SPM, based on the research topics, have been utilized: STM, cAFM, and ESM. In addition to the measured topography, STM/S could also reveal local electronic properties, in particular the electronic LDOS; cAFM can probe the local conductance profiles; and ESM could measure the local electrochemistry reactivity. The spatial resolutions are proven to be in the typical SPM resolution—10 nm down to atomic resolution depends on the techniques and the samples. With these capabilities, reported topics of interests using XSPM include: two-dimensional electron gas at LAO/STO interfaces; electron band alignment at LCMO/Nb:STO interfaces; lower electronic band gap at domain walls in BFO; band bending under different ferroelectric polarization configurations on BFO/Nb:STO; short range charge transfer between superconducting YBCO and ferromagnetic LCMO interfaces; Schottky electric field induced mechanical properties change; high interfacial electrochemical reactivity at BZY/NGO interfaces; LSM/STF interfaces with FIB; and oxygen vacancy-related electronic modification in LSC_{113}/LSC_{214} multilayers. All of these successful examples pioneered the way toward a brand new way of studying interfacial phenomena using various types of SPM in cross-sectional geometry. What is expected is the further expansion of using SPM in cross-sectional geometry for a wider variety of material systems, such as solar cell material systems [125, 126]. Combination of SPM with external stimuli, such as electric field, magnetic field, light illumination, mechanical stress, and so on are also expected to provide unprecedented views on the interfacial science. With the increasing interests in the novel interfacial phenomena found in complex oxide interfaces, XSPM will provide unique information toward understanding the underlying physics not accessible by other experimental techniques.

Acknowledgements TYC acknowledges the U.S. Department of Energy, Office of Basic Energy Sciences, Division of Materials Sciences and Engineering for financial support (DEFG02-10ER46728).

References

1. Ashcroft NW, Mermin ND (1976) Solid state physics. Saunders College, Philadelphia
2. Okamoto S, Millis AJ (2004) Electronic reconstruction at an interface between a Mott Insulator and a Band Insulator. Nature 428:630–633
3. Ramirez AP (1997) Colossal magnetoresistance. J Phys Condens Matter 9:8171–8199
4. Orenstein J, Millis AJ (2000) Advances in the physics of high-temperature superconductivity. Science 288(5465):468–474
5. Tybell T, Ahn CH, Triscone J-M (1999) Ferroelectricity in thin perovskite films. Appl Phys Lett 75:856–858
6. Fong DD, Stephenson GB, Streiffer SK, Eastman JA, Auciello O, Fuoss PH, Thompson C (2004) Ferroelectricity in ultrathin perovskite films. Science 304:1650–1653
7. Yanase N, Abe K, Fukushima N, Kawakubo T (1999) Thickness dependence of ferroelectricity in heteroepitaxial $BaTiO_3$ thin film capacitors. Jpn J Appl Phys 38:5305–5308
8. Chu Y-H, Martin LW, Holcomb MB, Ramesh R (2007) Controlling magnetism with multiferroics. Mater Today 10:16–23
9. Binnig G, Rohrer H, Gerber C, Weibel E (1983) 7 × 7 reconstruction on Si(111) resolved in real space. Phys Rev Lett 50:120–123
10. Moore RG, Zhang J, Nascimento VB, Jin R, Guo J, Wang GT, Fang Z, Mandrus D, Plummer EW (2007) A surface-tailored, purely electronic, Mott metal-to-insulator transition. Science 318:615–619
11. Parkin SSP, Sigsbee R, Felici R, Felcher GP (1986) Observation of magnetic dead layers at the surface of iron oxide films. Appl Phys Lett 48:604–606
12. Nascimento VB, Freeland JW, Saniz R, Moore RG, Mazur D, Liu H, Pan MH, Rundgren J, Gray KE, Rosenberg RA, Zheng H, Mitchell JF, Freeman AJ, Veltruska K, Plummer EW (2009) Surface-stabilized nonferromagnetic ordering of a layered ferromagnetic manganite. Phys Rev Lett 103:227201
13. van der Laan G, Hoyland MA, Surman M, Flipse CFJ, Thole BT (1992) Surface orbital magnetic moment of ferromagnetic nickel studied by magnetic circular dichroism in Ni 3p core level photoemission. Phys Rev Lett 69:3827–3830

14. Chakhalian J, Freeland JW, Srajer G, Strempfer J, Khaliullin G, Cezar JC, Charlton T, Dalgliesh R, Bernhard C, Cristiani G, Habermeier H-U, Keimer B (2006) Magnetism at the interface between ferromagnetic and superconducting oxides. Nat Phys 2:244–248
15. Nakagawa N, Hwang HY, Muller DA (2006) Why some interfaces cannot be sharp. Nat Mater 5:204–209
16. Chakhalian J, Freeland JW, Habermeier H-U, Cristiani G, Khaliullin G, van Veenendaal M, Keimer B (2007) Orbital reconstruction and covalent bonding at an oxide interface. Science 318:1114–1117
17. Baiutti F, Christiani G, Logvenov G (2014) Towards precise defect control in layered oxide structures by using oxide molecular beam epitaxy. Beilstein J Nanotechnol 5:596–602
18. Eason R (2006) Pulsed laser deposition of thin flms: applications—led growth of functional materials. Wiley, Hoboken
19. Christen HM, Eres G (2008) Recent advances in pulsed-laser deposition of complex oxides. J Phys Condens Matter 20:264005
20. Schiffer P, Ramirez AP, Bao W, Cheong S-W (1995) Low temperature magnetoresistance and the magnetic phase diagram of $La_{1-x}Ca_xMnO_3$. Phys Rev Lett 75:3336–3339
21. Oka T, Nagaosa N (2005) Interfaces of correlated electron systems: proposed mechanism for colossal electroresistance. Phys Rev Lett 95:266403
22. Eskes H, Meinders MBJ, Sawatzky GA (1991) Anomalous transfer of spectral weight in doped strongly correlated systems. Phys Rev Lett 67:1035–1038
23. Chien T-Y, Liu J, Chakhalian J, Guisinger N, Freeland J (2010) Visualizing nanoscale electronic band alignment at the $La_{2/3}Ca_{1/3}MnO_3$/Nb:$SrTiO_3$ interface. Phys Rev B 82:041101(R)
24. Chien T-Y, Kourkoutis LF, Chakhalian J, Gray B, Kareev M, Guisinger NP, Muller DA, Freeland JW (2013) Visualizing short-range charge transfer at the interfaces between ferromagnetic and superconducting oxides. Nat Commun 4:2336
25. Chien T-Y, Chakhalian J, Freeland JW, Guisinger NP (2013) Cross-sectional scanning tunneling microscopy applied to complex oxide interfaces. Adv Funct Mater 23:2565–2575
26. Huang B-C, Chiu Y-P, Huang P-C, Wang W-C, Tra VT, Yang J-C, He Q, Lin J-Y, Chang C-S, Chu Y-H (2012) Mapping band alignment across complex oxide heterointerfaces. Phys Rev Lett 109:246807
27. Huang BC, Chen YT, Chiu YP, Huang YC, Yang JC, Chen YC, Chu YH (2012) Direct observation of ferroelectric polarization-modulated band bending at oxide interfaces. Appl Phys Lett 100:122903
28. Li L, Richter C, Mannhart J, Ashoori RC (2011) Coexistence of magnetic order and two-dimensional superconductivity at $LaAlO_3/SrTiO_3$ interfaces. Nat Phys 7:762–766
29. Torrance JB, Lacorre P, Nazzal AI, Ansaldo EJ, Niedermayer C (1992) Systematic study of insulator-metal transitions in perovskites $RNiO_3$ (R=Pr,Nd,Sm,Eu) due to closing of charge-transfer gap. Phys Rev B 45:8209–8212
30. Hemberger J, Krimmel A, Kurz T, von Nidda H-AK, Ivanov VY, Mukhin AA, Balbashov AM, Loidl A (2002) Structural, magnetic and electrical properties of single crystalline $La_{1-x}Sr_xMnO_3$ for $0.4 < x < 0.85$. Phys Rev B 66:94410
31. Paraskevopoulos M, Mayr F, Hemberger J, Loidl A, Heichele R, Maurer D, Muller V, Mukhin AA, Balbashov AM (2000) Magnetic properties and the phase diagram of $La_{1-x}Sr_xMnO_3$ for $x \leq 0.2$. J Phys Condens Matter 12:3993
32. Paraskevopoulos M, Mayr F, Hartinger C, Pimenov A, Hemberger J, Lunkenheimer P, Loidl A, Mukhin AA, Ivanov VY, Balbashov AM (2000) The phase diagram and optical properties of $La_{1-x}Sr_xMnO_3$ for $x \leq 0.2$. J Magn Magn Mater 211:118–127
33. Ando Y, Komiya S, Segawa K, Ono S, Kurita Y (2004) Electronic phase diagram of High-T_C, cuprate superconductors from a mapping of the in-plane resistivity curvature. Phys Rev Lett 93:267001
34. Tomio T, Miki H, Tabata H, Kawai T, Kawai S (1994) Control of electrical conductivity in laser deposited $SrTiO_3$ thin films with Nb doping. J Appl Phys 76:5886–5890
35. Muller DA, Nakagawa N, Ohtomo A, Grazul JL, Hwang HY (2004) Atomic-scale imaging of nanoengineered oxygen vacancy profile in $SrTiO_3$. Nature 430:657–661

36. Choi M, Oba F, Kumagai Y, Tanaka I (2013) Anti-ferrodistortive-like oxygen-octahedron rotation induced by the oxygen vacancy in cubic $SrTiO_3$. Adv Mater 25:86–90
37. Lee JH, Fang L, Vlahos E, Ke X, Jung YW, Kourkoutis LF, Kim J-W, Ryan PJ, Heeg T, Roeckerath M, Goian V, Bernhagen M, Uecker R, Hammel PC, Rabe KM, Kamba S, Schubert J, Freeland JW, Muller DA, Fennie CJ, Schiffer P, Gopalan V, Johnston-Halperin E, Schlom DG (2010) A strong ferroelectric ferromagnet created by means of spin-lattice coupling. Nature 466:954–958
38. Meevasana W, King PDC, He RH, Mo S, Hashimoto M, Tamai A, Songsiriritthigul P, Baumberger F, Shen Z (2011) Creation and control of a two-dimensional electron liquid at the bare $SrTiO_3$ surface. Nat Mater 10:114–118
39. Tufte ON, Chapman PW (1967) Electron mobility in semiconducting strontium titanate. Phys Rev 155:796–802
40. Ohtomo A, Hwang HY (2004) A high-mobility electron gas at the $LaAlO_3/SrTiO_3$ heterointerface. Nature 427:423–426
41. Reyren N, Thiel S, Caviglia AD, Kourkoutis LF, Hammerl G, Richter C, Schneider CW, Kopp T, Ruetschi A-S, Jaccard D, Gabay M, Muller DA, Triscone J-M, Mannhart J (2007) Superconducting interfaces between insulating oxides. Science 317:1196–1199
42. Gozar A, Logvenov G, Kourkoutis LF, Bollinger AT, Giannuzzi LA, Muller DA, Bozovic I (2008) High-temperature interface superconductivity between metallic and insulating copper oxides. Nature 455:782–785
43. Rose V, Wang K, Chien T, Hiller J, Rosenmann D, Freeland JW, Preissner C, Hla S-W (2013) Synchrotron X-ray scanning tunneling microscopy: fingerprinting near to far field transitions on Cu(111) induced by synchrotron radiation. Adv Funct Mater 23:2646–2652
44. Wang K, Rosenmann D, Holt M, Winarski R, Hla S-W, Rose V (2013) An easy-to-implement filter for separating photo-excited signals from topography in scanning tunneling microscopy. Rev Sci Instrum 84:63704
45. Rose V, Freeland JW, Garrett R, Gentle I, Nugent K, Wilkins S (2010) Nanoscale chemical imaging using synchrotron X-ray enhanced scanning tunneling microscopy. AIP Conf Proc 1234:445–448
46. Cummings ML, Chien T, Preissner C, Madhavan V, Diesing D, Bode M, Freeland JW, Rose V (2012) Combining scanning tunneling microscopy and synchrotron radiation for high-resolution imaging and spectroscopy with chemical, electronic, and magnetic contrast. Ultramicroscopy 112:22–31
47. Rose V, Chien TY, Freeland JW, Rosenmann D, Hiller J, Metlushko V (2012) Spin-dependent synchrotron X-ray excitations studied by scanning tunneling microscopy. J Appl Phys 111:07E304
48. Rose V, Chien TY, Hiller J, Rosenmann D, Winarski RP (2011) X-ray nanotomography of SiO_2-coated $Pt_{90}Ir_{10}$ tips with sub-micron conducting apex. Appl Phys Lett 99:173102
49. Rose V, Freeland JW, Gray KE, Streiffer SK (2008) X-ray-excited photoelectron detection using a scanning tunneling microscope. Appl Phys Lett 92:193510
50. Bonnell DA (1998) Scanning tunneling microscopy and spectroscopy of oxide surfaces. Prog Surf Sci 57:187–252
51. Bonnell DA, Garra J (2008) Scanning probe microscopy of oxide surfaces: atomic structure and properties. Rep Prog Phys 71:44501
52. Chiu Y-P, Chen Y-CY-T, Huang B-C, Shih M-C, Yang J-C, He Q, Liang C-W, Seidel J, Chen Y-CY-T, Ramesh R, Chu Y-H (2011) Atomic-scale evolution of local electronic structure across multiferroic domain walls. Adv Mater 23:1530–1534
53. Basletic M, Maurice J-L, Carrétéro C, Herranz G, Copie O, Bibes M, Jacquet E, Bouzehouane K, Fusil S, Barthélémy A (2008) Mapping the spatial distribution of charge carriers in $LaAlO_3/SrTiO_3$ heterostructures. Nat Mater 7:621–625
54. Copie O, Garcia V, Bödefeld C, Carrétéro C, Bibes M, Herranz G, Jacquet E, Maurice J-L, Vinter B, Fusil S, Bouzehouane K, Jaffrès H, Barthélémy A (2009) Towards two-dimensional metallic behavior at $LaAlO_3/SrTiO_3$ interfaces. Phys Rev Lett 102:216804

55. Kuru Y, Jalili H, Cai Z, Yildiz B, Tuller HL (2011) Direct probing of nanodimensioned oxide multilayers with the aid of focused ion beam milling. Adv Mater 23:4543–4548
56. Kermode JR, Albaret T, Sherman D, Bernstein N, Gumbsch P, Payne MC, Csányi G, De Vita A (2008) Low-speed fracture instabilities in a brittle crystal. Nature 455:1224–1227
57. Ahmad AL, Idrus NF, Othman MR (2005) Preparation of perovskite alumina ceramic membrane using sol-gel method. J Membr Sci 262:129
58. Chien T, Guisinger NP, Freeland JW (2010) Survey of fractured $SrTiO_3$ surfaces: from the micrometer to nanometer scale. J Vac Sci Technol B Microelectron Nanometer Struct 28:C5A11–C5A13
59. Guisinger NP, Santos TS, Guest JR, Chien T-Y, Bhattacharya A, Freeland JW, Bode M (2009) Nanometer-scale striped surface terminations on fractured $SrTiO_3$ surfaces. ACS Nano 3:4132
60. Chien T, Santos TS, Bode M, Guisinger NP, Freeland JW (2009) Controllable local modification of fractured Nb-doped $SrTiO_3$ surfaces. Appl Phys Lett 95:163107
61. Chien T-Y, Guisinger NP, Freeland JW (2011) Cross-sectional scanning tunneling microscopy for complex oxide interfaces. Proc SPIE 7940:79400T
62. Chien T, Liu J, Yost AJ, Chakhalian J, Freeland JW, Guisinger NP (2016) Built-in electric field induced mechanical property change at the lanthanum nickelate/Nb-doped strontium titanate interfaces. Sci Rep 6:19017
63. Chen Y, Cai Z, Kuru Y, Ma W, Tuller HL, Yildiz B (2013) Electronic activation of cathode superlattices at elevated temperatures-source of markedly accelerated oxygen reduction kinetics. Adv Energy Mater 3:1221–1229
64. Chien T, Freeland JW, Guisinger NP (2012) Morphology control of Fe films using ordered termination on $SrTiO_3$ surfaces. Appl Phys Lett 100:31601
65. S. P. Mcjijnkins and J. I. Thornton, Glass fracture analysis. A review Forensic Sci 2, 1–27 (1973)
66. Reihl B, Bednorz JG, Müller KA, Jugnet Y, Landgren G, Morar JF (1984) Electronic structure of strontium titanate. Phys Rev B 30:803–806
67. Kohiki S, Arai M, Yoshikawa H, Fukushima S, Oku M, Waseda Y (2000) Energy-loss structure in core-level photoemission satellites. Phys Rev B 62:7964–7969
68. Pennec Y, Ingle NJC, Elfimov IS, Varene E, Maeno Y, Damascelli A, Barth JV (2008) Cleaving-temperature dependence of layered-oxide surfaces. Phys Rev Lett 101:216103
69. Lytle FW (1964) X-ray diffractometry of low-temperature phase transformations in strontium titanate. J Appl Phys 35:2212–2215
70. Herranz G, Copie O, Gentils A, Tafra E, Basletić M, Fortuna F, Bouzehouane K, Fusil S, Jacquet É, Carrétéro C, Bibes M, Hamzić A, Barthélémy A (2010) Vacancy defect and carrier distributions in the high mobility electron gas formed at ion-irradiated $SrTiO_3$ surfaces. J Appl Phys 107:103704
71. Lü WM, Sun JR, Wang DJ, Xie YW, Liang S, Chen YZ, Shen BG (2008) Interfacial potential in $La_{1-x}Ca_xMnO_3/SrTiO_3$:Nb junctions with different Ca contents. Appl Phys Lett 92:62503
72. Siemons W, Koster G, Yamamoto H, Harrison WA, Lucovsky G, Geballe TH, Blank DHA, Beasley MR (2007) Origin of charge density at $LaAlO_3$ on $SrTiO_3$ heterointerfaces: possibility of intrinsic doping. Phys Rev Lett 98:196802
73. Bert JA, Kalisky B, Bell C, Kim M, Hikita Y, Hwang HY, Moler KA (2011) Direct imaging of the coexistence of ferromagnetism and superconductivity at the $LaAlO_3/SrTiO_3$ interface. Nat Phys 7:767–771
74. Pojani A, Finocchi F, Noguera C (1999) Polarity on the $SrTiO_3$ (111) and (110) surfaces. Surf Sci 442:179–198
75. Herranz G, Sánchez F, Dix N, Scigaj M, Fontcuberta J (2012) High mobility conduction at (110) and (111) $LaAlO_3/SrTiO_3$ interfaces. Sci Rep 2:758
76. Willmott PR, Pauli SA, Herger R, Schlepütz CM, Martoccia D, Patterson BD, Delley B, Clarke R, Kumah D, Cionca C, Yacoby Y (2007) Structural basis for the conducting interface between $LaAlO_3$ and $SrTiO_3$. Phys Rev Lett 99:155502

77. Chambers SA, Engelhard MH, Shutthanandan V, Zhu Z, Droubay TC, Qiao L, Sushko PV, Feng T, Lee HD, Gustafsson T, Garfunkel E, Shah AB, Zuo JM, Ramasse QM (2010) Instability, intermixing and electronic structure at the epitaxial $LaAlO_3/SrTiO_3$ (001) heterojunction. Surf Sci Rep 65:317–352
78. Cen C, Thiel S, Hammerl G, Schneider CW, Andersen KE, Hellberg CS, Mannhart J, Levy J (2008) Nanoscale control of an interfacial metal-insulator transition at room temperature. Nat Mater 7:298–302
79. Zhao T, Scholl A, Zavaliche F, Lee K, Barry M, Doran A, Cruz MP, Chu YH, Ederer C, Spaldin NA, Das RR, Kim DM, Baek SH, Eom CB, Ramesh R (2006) Electrical control of antiferromagnetic domains in multiferroic $BiFeO_3$ films at room temperature. Nat Mater 5:823–829
80. Sone K, Naganuma H, Ito M, Miyazaki T, Nakajima T, Okamura S (2015) 100-nm-sized magnetic domain reversal by the magneto-electric effect in self-assembled $BiFeO_3/CoFe_2O_4$ bilayer films. Sci Rep 5:9348
81. Hsieh Y-H, Strelcov E, Liou J-M, Shen C-Y, Chen Y-C, Kalinin SV, Chu Y-H (2013) Electrical modulation of the local conduction at oxide tubular interfaces. ACS Nano 7:8627–8633
82. Spaldin NA, Fiebig M (2005) The renaissance of magnetoelectric multiferroics. Science 309:391–392
83. Streiffer SK, Parker CB, Romanov AE, Lefevre MJ, Zhao L, Speck JS, Pompe W, Foster CM, Bai GR (1998) Domain patterns in epitaxial rhombohedral ferroelectric films. J Appl Phys 83:2742–2753
84. Mermin ND (1979) The topological theory of defects in ordered media. Rev Mod Phys 51:591–648
85. Tokunaga Y, Furukawa N, Sakai H, Taguchi Y, Arima T, Tokura Y (2009) Composite domain walls in a multiferroic perovskite ferrite. Nat Mater 8:558–562
86. Béa H, Bibes M, Ott F, Dupé B, Zhu XH, Petit S, Fusil S, Deranlot C, Bouzehouane K, Barthélémy A (2008) Mechanisms of exchange bias with multiferroic $BiFeO_3$ epitaxial thin films. Phys Rev Lett 100:17204
87. Martin LW, Chu Y-H, Holcomb MB, Huijben M, Yu P, Han S-J, Lee D, Wang SX, Ramesh R (2008) Nanoscale control of exchange bias with $BiFeO_3$ thin films. Nano Lett 8:2050–2055
88. Lubk A, Gemming S, Spaldin NA (2009) First-principles study of ferroelectric domain walls in multiferroic bismuth ferrite. Phys Rev B 80:104110
89. Seidel J, Martin LW, He Q, Zhan Q, Chu Y-H, Rother A, Hawkridge ME, Maksymovych P, Yu P, Gajek M, Balke N, Kalinin SV, Gemming S, Wang F, Catalan G, Scott JF, Spaldin NA, Orenstein J, Ramesh R (2009) Conduction at domain walls in oxide multiferroics. Nat Mater 8:229–234
90. Buzdin AI (2005) Proximity effects in superconductor-ferromagnet heterostructures. Rev Mod Phys 77:935–976
91. Dagotto E (2005) Complexity in strongly correlated electronic systems. Science 309:257–262
92. Bibes M, Villegas JE, Barthélémy A (2011) Ultrathin oxide films and interfaces for electronics and spintronics. Adv Phys 60:5–84
93. Hoffmann A, Te Velthuis SGE, Sefrioui Z, Santamaría J, Fitzsimmons MR, Park S, Varela M (2005) Suppressed magnetization in $La_{0.7}Ca_{0.3}MnO_3/YBa_2Cu_3O_{7-\delta}$ superlattices. Phys Rev B 72:140407(R)
94. Visani C, Tornos J, Nemes NM, Rocci M, Leon C, Santamaria J, Te Velthuis SGE, Liu Y, Hoffmann A, Freeland JW, Garcia-Hernandez M, Fitzsimmons MR, Kirby BJ, Varela M, Pennycook SJ (2011) Symmetrical interfacial reconstruction and magnetism in $La_{0.7}Ca_{0.3}MnO_3/YBa_2Cu_3O_7/La_{0.7}Ca_{0.3}MnO_3$ heterostructures. Phys Rev B 84:060405(R)
95. Zhang ZL, Kaiser U, Soltan S, Habermeier H-U, Keimer B (2009) Magnetic properties and atomic structure of $La_{2/3}Ca_{1/3}MnO_3$-$YBa_2Cu_3O_7$ heterointerfaces. Appl Phys Lett 95:242505
96. Varela M, Lupini AR, Pennycook SJ, Sefrioui Z, Santamaria J (2003) Nanoscale analysis of $YBa_2Cu_3O_{7-x}/La_{0.67}Ca_{0.33}MnO_3$ interfaces. Solid State Electron 47:2245–2248

97. Liu B, Chen Z, Wang Y, Wang X (2001) The effect of an electric field on the mechanical properties and microstructure of Al–Li alloy containing Ce. Mater Sci Eng A A313:69–74
98. Liu W, Liang KM, Zheng YK, Cui JZ (1996) Effect of an electric field during solution treatment of 2091 Al-Li alloy. J Mater Sci Lett 15:1327–1329
99. Conrad H, Guo Z, Sprecher AF (1989) Effect of an electric field on the recovery and recrystallization of Al and Cu. Scr Metall 23:821–823
100. Sprecher AF, Mannan SL, Conrad H (1986) On the mechanisms for the electroplastic effect in metals. Acta Metall 34:1145–1162
101. Park Y, Kim HG (1997) Effect of electric field on the phase transition in $ZrTiO_4$. J Mater Sci Lett 16:1130–1132
102. Kumar S, Singh RN (1997) Influence of applied electric field and mechanical boundary condition on the stress distribution at the crack tip in piezoelectric materials. Mater Sci Eng A 231:1–9
103. Yang DI, Conrad H (1998) Influence of an electric field on the plastic deformation of polycrystalline NaCl at elevated temperatures. Acta Mater 46:1963–1968
104. Yang F, Dang H (2009) Effect of electric field on the nanoindentation of zinc sulfide. J Appl Phys 105:56110
105. Revilla RI, Li X-J, Yang Y-L, Wang C (2014) Large electric field-enhanced-hardness effect in a SiO_2 film. Sci Rep 4:4523
106. Wang H, Lin H-T, Wereszczak AA (2010) Strength properties of poled lead zirconate titanate subjected to biaxial flexural loading in high electric field. J Am Ceram Soc 93:2843–2849
107. Yamamoto T, Suzuki S, Kawaguchi K, Takahashi K (1998) Temperature dependence of the ideality factor of $Ba_{1-x}K_xBiO_3$/Nb-doped $SrTiO_3$ all-oxide-type Schottky junctions. Jpn J Appl Phys 37:4737–4746
108. Stengel M, Spaldin NA (2006) Origin of the dielectric dead layer in nanoscale capacitors. Nature 443:679–682
109. Ryu KH, Haile SM (1999) Chemical stability and proton conductivity of doped $BaCeO_3$-$BaZrO_3$ solid solutions. Solid State Ionics 125:355–367
110. Malagoli M, Liu ML, Park HC, Bongiorno A (2013) Protons crossing triple phase boundaries based on a metal catalyst, Pd or Ni, and barium zirconate. Phys Chem Chem Phys 15:12525–12529
111. Pergolesi D, Fabbri E, D'Epifanio A, Di Bartolomeo E, Tebano A, Sanna S, Licoccia S, Balestrino G, Traversa E (2010) High proton conduction in grain-boundary-free yttrium-doped barium zirconate films grown by pulsed laser deposition. Nat Mater 9:846–852
112. Foglietti V, Yang N, Tebano A, Aruta C, Di Bartolomeo E, Licoccia S, Cantoni C, Balestrino G (2014) Heavily strained $BaZr_{0.8}Y_{0.2}O_{3-x}$ interfaces with enhanced transport properties. Appl Phys Lett 104:81612
113. Yang N, Cantoni C, Foglietti V, Tebano A, Belianinov A, Strelcov E, Jesse S, Di Castro D, Di Bartolomeo E, Licoccia S, Kalinin SV, Balestrino G, Aruta C (2015) Defective interfaces in yttrium-doped barium zirconate films and consequences on proton conduction. Nano Lett 15:2343–2349
114. Kumar A, Ciucci F, Morozovska AN, Kalinin SV, Jesse S (2011) Measuring oxygen reduction/evolution reactions on the nanoscale. Nat Chem 3:707–713
115. Kumar A, Ciucci F, Leonard D, Jesse S, Biegalski M, Christen H, Mutoro E, Crumlin E, Shao-Horn Y, Borisevich A, Kalinin SV (2013) Probing bias-dependent electrochemical gas-solid reactions in $(La_xSr_{1-x})CoO_{3-\delta}$ cathode materials. Adv Funct Mater 23:5027–5036
116. Dholabhai P, Pilania G, Aguiar J, Misra A, Uberuaga BP (2014) Termination chemistry-driven dislocation structure at $SrTiO_3$/MgO heterointerfaces. Nat Commun 5:5043
117. Katsiev K, Yildiz B, Balasubramaniam KR, Salvador P (2009) Electron tunneling characteristics on $La_{0.7}Sr_{0.3}MnO_3$ thin-film surfaces at high temperature. Appl Phys Lett 95:92106
118. Jung W, Tuller HL (2008) Investigation of cathode behavior of model thin-film $SrTi_{1-x}Fe_xO_{3-\delta}$ (x=0.35 and 0.5) mixed ionic-electronic conducting electrodes. J Electrochem Soc 155:B1194–B1201

119. Sase M, Hermes F, Yashiro K, Sato K, Mizusaki J, Kawada T, Sakai N, Yokokawa H (2008) Enhancement of oxygen surface exchange at the hetero-interface of (La,Sr)CoO_3/(La,Sr)$_2$$CoO_4$ with PLD-layered films. J Electrochem Soc 155:B793–B797
120. Sase M, Yashiro K, Sato K, Mizusaki J, Kawada T, Sakai N, Yamaji K, Horita T, Yokokawa H (2008) Enhancement of oxygen exchange at the hetero interface of (La,Sr)CoO_3/(La,Sr)$_2$$CoO_4$ in composite ceramics. Solid State Ionics 178:1843–1852
121. Januschewsky J, Ahrens M, Opitz A, Kubel F, Fleig J (2009) Optimized $La_{0.6}Sr_{0.4}CoO_{3-\delta}$ thin-film electrodes with extremely fast oxygen-reduction kinetics. Adv Funct Mater 19:3151–3156
122. Adler SB (1998) Mechanism and kinetics of oxygen reduction on porous $La_{1-x}Sr_xCoO_{3-\delta}$ electrodes. Solid State Ionics 111:125–134
123. Mastrikov YA, Merkle R, Heifets E, Kotomin EA, Maier J (2010) Pathways for oxygen incorporation in mixed conducting perovskites: a DFT-based mechanistic analysis for (La,Sr)$MnO_{3-\delta}$. J Phys Chem C 114:3017–3027
124. Han JW, Yildiz B (2012) Mechanism for enhanced oxygen reduction kinetics at the (La,Sr)$CoO_{3-\delta}$/(La,Sr)$_2$$CoO_{4+\delta}$ hetero-interface. Energy Environ Sci 5:8598–8607
125. Yost AJ, Pimachev A, Ho C-C, Darling SB, Wang L, Su W-F, Dahnovsky Y, Chien T (2016) Coexistence of two electronic nano-phases on a $CH_3NH_3PbI_{3-x}Cl_x$ surface observed in STM measurements. ACS Appl Mater Interfaces 8:29110–29116
126. Shih M-C, Huang B-C, Lin C-C, Li S-S, Chen H-A, Chiu Y-P, Chen C-W (2013) Atomic-scale interfacial band mapping across vertically phased-separated polymer/fullerene hybrid solar cells. Nano Lett 13:2387–2392

Chapter 6
A Review of Nanofluid Synthesis

Binjian Ma and Debjyoti Banerjee

6.1 Introduction

Nanofluids are stable colloidal suspensions of nanoparticles in a chosen solvent (base fluid). The dimension of "nano-" entities is strictly defined by ISO/TS 27687 as a particle with at least one representative dimension (e.g., diameter, thickness, or length) ranging in size from 1 to 100 nm. This conventional definition of the nanoparticle is commonly agreed among nanotechnology researchers. The classification of dimension less than 100 nm arises from the phenomena that the surface area to volume ratio increases rapidly as the size of the particle is diminished. As a result, the material properties of these novel nanoparticles often deviate significantly and anomalously from their bulk (conventional) values due to the dominance of surface effects (e.g., from interfacial interactions and dominant contributions from surface energy). In other words, nanoparticles exhibit unique properties which are drastically different from their bulk characteristics. In-depth studies on the transport mechanisms responsible for these anomalous behaviors are still a burgeoning topic in contemporary literature.

The concept of nanofluid was first proposed by Choi and Eastman [1] in 1995. They observed an anomalous enhancement in the thermal conductivity when a small percentage of copper nanoparticles were dispersed in water. Subsequently, various combinations of nanoparticles and liquids have been studied for enhancing

B. Ma
Department of Mechanical Engineering, Texas A&M University,
College Station, TX 77843-3123, USA

D. Banerjee (✉)
Department of Mechanical Engineering, Texas A&M University,
College Station, TX 77843-3123, USA

Department of Petroleum Engineering (Joint Courtesy Appointment), Mary Kay O'Connor
Process Safety Center, Texas A&M University, College Station, TX 77843-3123, USA
e-mail: dbanerjee@tamu.edu

G. Balasubramanian (ed.), *Advances in Nanomaterials*,
DOI 10.1007/978-3-319-64717-3_6

transport phenomena and thermophysical properties (e.g., thermal conductivity, specific heat capacity, and viscosity). Typical materials used for synthesizing nanofluids are metals (copper, aluminum, gold, etc.), inorganic oxides (iron oxide, zinc oxide, silicon dioxide, etc.), carbon-based materials (CNT, graphene, fullerene, etc.), and other ceramics (aluminum nitride, PNP, cellulose, etc.). A popular choice for solvents includes water, glycols, ionic liquids, organic liquids, and refrigerants. Other liquids explored in the literature include lubricants, oils, biofluids, emulsions, fuels (e.g., kerosene), alcohols, molten salt eutectics, etc.

Despite the vast range of materials that have been explored in the literature for synthesizing nanofluids, the general concept of dispersing nanoparticles in the base fluid is to enhance certain material properties of the base fluid for achieving enhanced performance in a chosen application. For example, high thermal conductivity is desired in the heat transfer applications (often at the expense of higher viscosity and pump penalty) while improved load-carrying capacity and non-Newtonian rheological behavior are preferred in lubrication applications (with a concomitant ability to dissipate heat rapidly). On the other hand, pump penalty is of secondary consideration in thermal energy storage (TES) applications—where the material cost ($\$/kWh_t$) and enhanced specific heat capacity are of primary importance. Hence the desired fluid properties are motivated by the chosen application which in turn mediates the choice of the synthesis protocol for a specific nanofluid. Thus, it is crucial to establish a library of synthesis protocols for the select set of nanofluids for target properties (and applications).

Another important issue and common concern in nanofluids application is the long-term stability. The current architecture in nano-manufacturing is well-developed which allows us to synthesize nanoparticles with different size, shape, and structure. However, keeping nanoparticle suspended uniformly in the base fluid for long enough time is still a challenging task. Various approaches have been explored to increase the stability of suspensions. These approaches have largely yielded short-term improvement but no conclusive studies exist for ensuring long-term stability. One key idea in keeping nanoparticle suspensions stable for long time periods is to prevent them from agglomerating (which is primarily induced by Brownian motion-mediated collision and coagulation). Some of the typical strategies utilized for ensuring long-term stability will be discussed in subsequent sections.

One last major consideration in nanofluid synthesis is the cost ($\$/kg$ and $\$/kWh_t$) and feasibility of scaling-up synthesis to large volumes. Such topics of engineering significance are often neglected in fundamental research but are key to the commercial success of nanofluids, if this technology were to penetrate conventional and novel engineering applications. The engineering economics of nanosynthesis, scale-up, and cost will therefore be explored briefly in this review.

6.2 Nanofluid Synthesis Protocols

Nanofluids are typically synthesized by either two-step method or one-step method.

The two-step method, as the name suggests, consists of two separate processes for synthesizing nanoparticles (or commercially procured) and dispersing the procured nanoparticles into base fluid. Such methods are being extensively used in nanofluid research due to its simplicity. Proper choice of synthesis protocol can enable better stability as well as control over precision and size of nanoparticles in the suspension. Nanomaterials used in this method are usually procured commercially, typically in the form of dry powders. With advances in synthesis techniques for nanoparticles, large-scale production of nanoparticles with good precision in size and shape has been achieved by commercial vendors (e.g., using combustion synthesis techniques). The techniques used for making nanoparticles include mechanical methods (milling, grinding, etc.), physical methods (physical vapor deposition, inert gas condensation, etc.), and chemical synthesis (sol-gel process, solution combustion, electrolysis, combustion synthesis, etc.). Depending on the requirements of the chosen applications, different synthesis protocols can be selected to deliver the specification for nanomaterials with the desired constraints for size and shape. This review will be limited to discussion of dispersion protocols rather than the synthesis of nanoparticles in the form of dry powders. Excellent reviews are available in the literature on the topic of the nanoparticle synthesis (e.g., by C. N. R. Rao [2]). These synthesis protocols for nanoparticles in the form of dry powders are categorized into: (a) top-down, and (b) bottom-up techniques. The reader interested in this topic can consult this reference (and similar reviews available in the literature).

The primary bottleneck of the two-step method is that appropriate dispersing technique is needed to ensure the stability of the nanoparticle suspensions in the solvent/fluid phase. Due to the high surface energy nanoparticles inherently form unstable suspensions in liquids and have a strong tendency to agglomerate (which is primarily caused by Brownian motion mediated collision of nanoparticles in the suspension). Various strategies are used for mitigating mutual collision of nanoparticles in the suspension and preventing (or minimizing) the tendency for agglomeration. Typical methods used for preventing agglomeration of nanoparticles include ultrasonic agitation, additives to control pH (e.g., buffer solutions), and functionalization of the nanoparticle with surfactants or chemical groups (such as amines or silanes) to promote better dissolution in the solvent phase by hydrogen bonding and/or ionic interactions. Ultrasonic agitation often provides limited payback since the nanoparticle suspensions typically settle and precipitate from the solution under gravity. The remaining approaches are intrusive methods since they often cause significant alteration of the chemical composition of the solvent phase. The details of nanofluid stabilization will be discussed in the later section. However, the two-step method often fails to deliver long-term stability of the synthesized nanofluids.

The one-step method relies on the generation of nanoparticles in-situ in the solvent phase from precursors. In other words, the synthesis and dispersion of nanoparticles happen simultaneously in the solvent phase. As a result, the propensity of agglomeration of the nanoparticles generated in-situ is minimized. The one-step method can be implemented by either a physical technique (e.g., direct evaporation and condensation) or chemical technique (e.g., chemical decomposition). However, it is more difficult to control the morphology of the particles precisely as small variations in the designed synthesis conditions (temperature, time, feeding rate, etc.) can drastically alter the properties of the synthesized nanofluids due to variation in nanoparticle size distribution and stability. Thus, it is very important to understand, model, and optimize the synthesis conditions to enable better control over the transport mechanisms (e.g., homogeneous or heterogeneous nucleation of the nanoparticles from the precursors as well as growth and subsequent agglomeration of the nanoparticles generated in-situ). It should be noted that if the generation of the nanoparticle is a distinctly separate process from the dispersion step, such methods are categorized as a two-step process.

6.2.1 Two-Step Method

For most nanofluid synthesized via two-step method, commercial nano-powders supplied by manufacturers were used directly. The preparation process itself is rather straightforward: the nanoparticles are first dispersed in the base fluid and then stabilized by different approaches. However, depending on the material of the nanoparticle and base fluid, the dispersion process could be either "spontaneous" or "non-spontaneous". In the former case, the nanoparticles would readily spread out in the base fluid and remain in the stable suspension state, while in the latter case, the nanoparticles tend to stay together unless external forces are applied. Such variation gives rise to the difference in the nanofluid preparation procedure.

In general, the nanofluids tend to be more stable if there is a strong affinity between the nanoparticles and liquid molecules and a strong repulsive force between nanoparticles. A good example for illustrating particle–solvent interaction would be the dispersion of TiO_2 in water. Due to the proximity of acidic and basic sites on the surface of different TiO_2 crystals (with/without defects), water molecules are found to get absorbed and dissociated on TiO_2 surfaces, with hydroxyl groups binding to surface Ti atom and H atoms binding with the bridging O atom [3–6]. Such feature allows the TiO_2 nanoparticles to form strong bonding with the water solvent, which reduces their possibility of coagulation in the suspension and enhances the stability of the nanofluid consequently. The dispersion of CNTs in water, on the contrary, goes to another extreme. Carbon by nature is almost purely aromatic and non-polar. They possess strong van der Waal forces between each other and high level of hydrophobicity. Thus, CNTs have a strong tendency to form agglomerates with the neighboring groups in common organic and aqueous media unless coated with some stabilizing agent/function group [7, 8].

Strong repulsive forces could also help keeping nanoparticle apart from each other, thereby increasing their stability in suspensions. Such forces could come from the electric double layer (charged stabilization) or absorbed polymeric molecules (steric stabilization) on the particle surface. Adjusting pH is a typical method for enhancing nanofluid, as the higher concentration of H^+/OH^- in the system increases surface charge density, and thus brings higher electrostatic potentials between particles [9]. Adding appropriate ionic compound could induce similar sterilization effect, as the ions transfer their charges to the nanoparticle surface and hence increase electrostatic stabilization [10].

The intermolecular potentials have a direct impact on the collision frequency of nanoparticles in the liquid environment. Apart from this, the stability of nanofluid is also dependent on the probability of merging upon collision, which is closely linked to the specific surface energy of the nanoparticles. From chemical thermodynamic point of view, a high specific surface energy is representative of an unstable state, in which the particles will try to minimize its free energy by forming large aggregates and reduce surface areas. For example, conducting (metal) nanoparticles possess very high specific energy of metal surfaces (1000~2000 mJ/m^2) in comparison with other organic and inorganic materials (~20 mJ/m^2 for teflon and 462 mJ/m^2 for silica) [11]. This results in high instability of "original" metallic nanofluids such that the freshly dispersed nanoparticles would agglomerate and precipitate readily in short time [12].

The point here is that the preparation of nanofluid via two-step method is more than simply mixing the particles with the fluid. Considerable efforts should be focused on how to obtain "stable" nanofluid with additional procedures before, during, and after the mixing process. Such practices are determined by the materials used in nanofluid preparation as well as the experimental conditions. Here, we will briefly summarize the two-step process used for preparing different nanofluids. More details on the different approaches for enhancing nanofluid stability will be discussed below. A general compilation of two-step nanofluid preparation can be found in the appendix for readers' reference.

Typical nanoparticles used in nanofluids are metallic (Cu, Ag, Au, Al, Fe), ceramic (CuO, Al_2O_3, Fe_3O_4, SiO_2, TiO_2, ZiO_2, AlN, SiN), or carbon-based (single/multi-walled CNT, graphene, C_{60}). The typical base fluid used include water/water-soluble molecular liquid (water, ethylene glycol, ethanol), oil (PAO, silicon oil, engine lubrication oil), and ionic liquid ([Bmim][PF_6], [HMIM]BF_4, molten salt). The table below summarizes the stability level of the corresponding nanofluids without any stabilizing agent/surface functionalization. It represents the inherent dispersibility of the nanoparticles in base fluids (Table 6.1).

Table 6.1 Stability of nanoparticles in different fluids

Without treatment	Water/water-soluble molecular liquid	Oil	Ionic liquid
Metallic	Poor	Moderate	Good
Ceramic	Moderate	Good	Very good
Carbon	Very poor	Poor	Moderate

6.2.2 One-Step Method

Although two-step method has been used widely in the nanofluid research community, the issue of agglomeration in the mixed nanofluid has promoted the exploration of the one-step method in which the synthesis and dispersion of nanoparticles are performed simultaneously. Depending on the nature of the synthesis approach, they could be classified into either physical or chemical method.

6.2.2.1 Physical Methods

The advantage of physical synthesis methods in comparison with chemical processes is the absence of solvent contamination in the process of nanoparticle generation. Evaporation-condensation is a typical physical approach for preparing nanofluids in one step, but other methods are also available.

Physical Vapor Deposition

Physical vapor deposition is a broad category of methods used for nanomaterial synthesis, in which nanoparticles were formed by direct condensation of the target metal vapor in contact with a flowing liquid. Different particle concentration and diameter can be achieved by controlling the vapor release and liquid flow rates. Such methods originated from the gas evaporation method used for preparing fine metal particles in the inert gas environment [13], and has been improved and adapted for producing monodispersed nanoparticles in the liquid environment. Akoh [14] synthesized ferromagnetic metal oxide (Fe_3O_4, CoO, and NiO)-based nanofluid by the so-called VEROS (Vacuum Evaporation onto a Running Oil Substrate) method in which particles were generated in oil with an average diameter of 2.5 nm. Wagner [15] prepared silver and iron nanofluids in oil using a similar approach involving magnetron sputtering. Eastman [16] prepared Cu/ethylene glycol nanofluid by evaporating the source metal into cooled liquid using resistive heating (10 nm Cu nanoparticles were produced in EG with 0.5% volume concentration and little agglomeration was observed).

The nanofluid synthesis via conventional PVD chamber requires low vapor pressure base fluid and high power for metal vaporization. Localized high temperature/heat flux technique can be used for achieving more efficient and convenient nanofluid synthesis. Exploding wire method (also known as pulsed wire explosion, pulsed-wire evaporation method) is one common technique used for creating metal vapor with localized high energy input [17]. Lee [18] used pulsed-wire evaporation technique for preparing ethylene glycol-ZnO nanofluid. In their experiment, a pulsed high-voltage DC power was used to induce a non-equilibrium overheating in a thin Zn wire. The pure metal evaporated within microseconds and condensed into small-size particles (<100 nm) after coming into direct contact with the EG. Park

[19] synthesized three different metallic nanofluids including Ag, Cu, and Al nanoparticles in three kinds of fluids: water, ethanol, and ethylene glycol using the electric explosion method. The author reported that higher energy leads to the decrease in the size of metallic nanoparticles. Similar methods were also used for preparing copper [20–22], silver [22–25], iron [22], gold [26], alumina [27], and titania [28] nanofluids.

Nanofluid synthesis with the aid of plasma is also a promising approach due to multiple advantages of this method—including the simplicity of the experimental design. The plasma could be generated either in the air or in the fluid. Chang [29] fabricated Al_2O_3 nanofluid by a modified plasma arc system. In the system, bulk metal was vaporized by high-temperature plasma arc and cooled by pre-condensed deionized water. The rapid cooling process prevented the growth of particle size which produced stable nanofluid with fine particles (25–75 nm). Teng [30] prepared organic nanofluid using a similar system, in which carbon was vaporized in a plasma chamber and cooled by deionized water to form fine carbon nanoparticles (244–284 nm). It should be noted that if pure metallic nanofluids are desired, the plasma should be generated in an inert gas environment to avoid the metal vapor from being oxidized.

Nanofluids can also be prepared directly by solution plasma in which the target material gets vaporized and condensed instantly. For submerged arc plasma methods, a selected metal is heated and vaporized by arc sparking between two electrodes immersed in dielectric liquids. The metallic aerosol then immediately condenses into nanoparticles under the cooling effect of the flowing liquid. Tsung [31] first used the arc-submersed system for synthesizing copper nanoparticle suspensions in de-ionized water (DIW). Cu nanoparticles with either coarse or fine bamboo leave structures (<200 nm) were generated by changing the environmental pressure during synthesis. The technique was adopted and improved for preparing stable TiO_2 [32], CuO/Cu_2O [33], silver [34], and nickel [35] nanoparticle suspension by the same research group, in which nanoparticles with different morphologies were manufactured by changing electric current, arc pulse-duration/off time and dielectric liquids. Saito and Akiyama [36] have made a thorough compilation of available nanomaterial synthesis techniques using solution plasma, where the source materials, reaction media, and electrode configurations are discussed in detail.

Laser ablation in the liquid is another simple and effective way of vaporizing metal solids and synthesizing nanofluids in one step. The method works by focusing a high-power laser at the submerged solid surface for an appropriate time until the solid melts and vaporizes above ablation threshold. Meanwhile, a thin liquid layer near the solid surface will also vaporize with the metal. The expansion of liquid and conversion to vapor fractures the metal into nano-sized drops, which are later supercooled by the surrounding liquid and transformed into nanoparticles [37]. Phuoc [38] synthesized Ag-deionized water nanofluid using multi-beam laser ablation (the synthesized nanoparticles were observed to be stable for several months). It was found that both laser intensities and multi-beam ablation can increase the ablation rate and promote the reduction of the nanoparticle size. Kim [39] prepared bare Au-water nanofluid using single-pulsed laser beam ($\lambda = 532$ nm). The average size of the nanoparticles ranged from 7.1 to 12.1 nm and the nanofluid samples were observed to be stable for 3 months after synthesis. The volume concentration of the synthesized nanofluids was 0.018%. The one-step laser ablation technique has also been used to synthesize a variety of nanofluids with different nanoparticles including: Cu [40], Al [41], Sn [42], Si/SiC [43], CuO [44], Al_2O_3 [45], and carbon particles [46, 47]. Compared to other methods, laser ablation in the liquid is a rather simple and "green" (environmentally friendly) technique for synthesizing nanofluids in water or other organic liquids under ambient conditions. More details on the fundamental mechanism and fabrication process of nanomaterials via laser ablation in liquid were reported by Zeng [48].

Wet Mechano-Chemical Techniques

The top-down approach for synthesizing nanoparticles via purely mechanical actuation (i.e., crushing, milling, and grinding) is an economical, simple, and environmentally benign alternative for synthesizing various nanofluids. With the extensive development and use of high-energy ball milling (HEBM), synthesis of ultrafine nanoparticles has been proven to work for a number of materials [49–51]. Inkyo [52] prepared a well-dispersed suspension of TiO_2 nanoparticles in methyl methacrylate (MMA) with 5% mass fraction using beads milling and centrifugal bead separation. Particle size distribution between 10 and 50 nm was achieved using 660 min milling time, and the nanoparticles remained in stable suspension with no sedimentation after 24 h. Harjanto [53] prepared TiO_2-water nanofluids through the one-step process in which titania nanoparticles were milled together with distilled water in a vial placed in a planetary ball mill. The concentrated nanofluid solution synthesized in this process was diluted into different concentrations and stabilized using ultrasonic stirrer and oleic acid served as a surfactant. The average nanoparticle size was in the range of 24.1–27.2 nm and good stability was confirmed from the measured values of absolute zeta potential. Nine [54] prepared well-dispersed Cu/Cu_2O-water nanofluid using low energy ball milling in aqueous solution by varying the ball size and milling period. Samples that have been ground for 30 min by 1 mm balls and for 60 min by 3 mm balls were reported as stable colloids after performing sedimentation tests for over 7 days. Almasy [55] prepared ferrofluids with particle size of 10–15 nm using vibrating ball mill with industrial magnetite powder. It was observed that the wet-milled magnetite suspension had a higher saturation magnetization than that obtained in the relatively rapid co-precipitation synthesis. Wet mechanochemical methods may be preferred in certain practical situations for preparing nanofluids due to their simplicity, but scaling-up these synthesis techniques for large-scale manufacturing is still challenging.

6.2.2.2 Chemical Methods

One-step synthesis of nanofluids via chemical methods involves the generation and dispersion of target nanoparticles using a single-step or a series of chemical reactions (i.e., reduction, decomposition, etc.) from certain precursors. The process is usually performed in liquid environments and is known as wet chemistry method. Compared to the physical methods, wet chemistry routines are generally cheaper, require minimal instrumentation and are easier to implement, especially for scaling-up these synthesis techniques for large-scale manufacturing as well as for mass production. Also, very precise control over monodispersed nanostructures is achievable by proper choice of the chemistry of the synthesis protocols.

Chemical Reduction Method

Chemical reduction method has been used for nanofluid preparation by a number of researchers. Liu [56] prepared Cu-water nanofluid using copper acetate as precursor and hydrazine (N_2H_4) as reducing agent. The nano-suspension with monodisperse Cu particles (50–100 nm) was formed by slowly adding a predetermined quantity of hydrazine solution into copper acetate solution by constant stirring. Garg [57] synthesized copper nanofluids in ethylene glycol using a similar method except for the addition of sodium hydroxide as an additional reactant. The mean nanoparticle size was measured to be 200 nm. Kumar [58] prepared stable non-agglomerated copper nanofluids by reducing copper sulfate pentahydrate with sodium hypophosphite as reducing agent in ethylene glycol as base fluid by means of conventional heating. The process was fast and cost-competitive. Shenoy [59] employed a different one-step reduction routine for preparing copper nanofluids, in which copper nitrate was reduced by glucose in the presence of sodium lauryl sulfate. The nanofluid was found to be stable for a minimum of 3 weeks, and the method was reported to be reliable, simple, and cost-effective.

Other types of nanofluids have also been synthesized using chemical reduction methods. Tsai [60] prepared aqueous gold nanofluid by the reduction of aqueous hydrogen tetrachloroaurate ($HAuCl_4$) with trisodium citrate and tannic acid. Au nanoparticles with different sizes were obtained by changing the amounts of tetrachloroaurate, trisodium citrate, and tannic acid. Xun [61] prepared stable silver nanofluid in kerosene by first extracting silver nitrate in nonpolar solvent by thio-substituted phosphinic acid extractant Cyanex 302, and then reducing Ag^+ solid. Salehi [62] prepared silver nanofluids using a different approach, in which silver nitrate was reduced by sodium borohydride and hydrazine using polyvinylpyrrolidone (PVP) as the surfactant. It should be noted that the chemical reduction method only works for a very limited number of metallic particles which are chemically inert.

Precipitation Method (Ion Exchange)

Nanoparticles can be synthesized in various liquids by interactions between different ions at a controlled rate. Cao [63] prepared 5 nm ZnO nanorods in ethanol by adding sodium hydroxide into zinc acetate dehydrate ($Zn(Ac)_2 \cdot 2H_2O$) solution at room temperature. It was found that the size and shape of nanorods with different size and shapes can be tuned via simply altering NaOH concentration and reaction time. Darezereshki [64] prepared maghemite (γ-Fe_2O_3) nanoparticles by mixing the $FeCl_3/FeCl_2 \cdot 4H_2O$ solution and the$NH_3 \cdot H_2O$ solution with vigorous stirring for 2 h. The average particle size was observed to be 45 nm. Manimaran [65] prepared copper oxide nanofluids by mixing copper chloride with sodium hydroxide in deionized water along with heating and magnetic stirring. The precipitates were observed to have an average size of 20 nm with very little agglomeration. Chakraborty [66] synthesized Cu-Al layered double hydroxide nanofluid via the one-step method. The nanoscale precipitation was formed by the dropwise addition of NaOH solution into the aqueous mixture solution of $Cu(NO_3)_2 \cdot 3H_2O$, $Al(NO_3)_3 \cdot 9H_2O$, and $NaNO_3$. The particles were observed to form clusters with size ranging from 86 to 126 nm in the suspension, and the nanofluid was found to be stable with high zeta potential value. Nanoparticles were also synthesized by the precipitation method first and re-dispersed in the fluid later after additional treatment (i.e., centrifugation, washing, calcination) [67–70]. Such methods should not be categorized as the one-step method as the synthesis and dispersion processes were not conducted simultaneously.

Sol-Gel Method (Hydrolysis)

Sol-gel process is a widely used technique for synthesizing nanoparticles with different size and scale of gel networks. The reaction mechanism involves two stages: (1) hydrolysis reaction of the precursor in which the functional binders of the precursors are substituted with the hydroxyl group; and (2) polycondensation reaction in which the hydroxyl group of monomers gets connected and forms continuous network [71]. Most nanofluid literature involving sol-gel process have adopted a two-step method for preparing the nanofluid samples, in which the nanoparticles were first generated from precursors, separated out from the reacting liquid, calcinated, and re-dispersed into the targeting base fluid [72–78]. However, there are also few cases where the nanofluid is prepared directly via the sol-gel method. Kim et al. [79] prepared stable silica nanofluid in water using TEOS as precursor, ethanol as solvent, DI water for hydrolysis reaction, and ammonium hydroxide as base catalyst. The particle size was approximately 30 nm, 70 nm, and 120 nm by controlling the ammonium hydroxide concentrations at 0.28 mol, 0.42 mol, and 0.56 mol, respectively. Jing [80] prepared highly dispersed silica-water nanofluid samples using a similar one-step sol gel process. Nanofluids with particle sizes of 5, 10, 25, and 50 nm were fabricated by first bending H_2O, NH_3 and alcohol, and then adding

TEOS with different quantities. The main issue with preparing nanofluid by sol-gel in one step is the contamination from excess reacting chemicals.

Emulsion-Polymerization

Emulsions are dispersed systems with two immiscible liquids (i.e., oil and water) in which liquid droplets are dispersed in a liquid medium. When micro/nano-emulsion materials are mixed, the reactant exchange from the dissolved molecules in each liquid could potentially promote precipitation reactions in the nanodroplets, which is then followed by nucleation, growth, and polymerization of the nanoparticles [81]. Han [82] developed a one-step, nanoemulsification method to synthesize the indium/polyalphaolefin (PAO) nanofluid. In the test, an indium pellet was first added to the PAO oil in a reaction vessel heated to 20 °C above the indium melting temperature. A PAO aminoester dispersant was injected into the reaction vessel, which also acts as a stabilizer to prevent nanoparticle coagulation. The emulsion was then exposed to high-intensity ultrasound radiation for more than 2 h until a stable nanofluid was formed. Kim [83] prepared biphasic nano-colloids of poly(dimethyl siloxane) (PDMS) and an organic copolymer (methyl acrylate co-methyl methacrylate co-vinyl acetate) in aqueous solution via emulsification and polymerization route. The particles were observed to have 170 nm size with a spatial resolution of 8 nm achieved in the STEM images. Pattekari [84] prepared stable nano-encapsulation of poorly soluble anticancer drug in water using a sonication assisted layer-by-layer polyelectrolyte coating (SLbL). In the experiment, polyanion solutions were added into the drug powder-contained DI water. The process involves simultaneous breaking down of the drug powder by ultrasonication and formation of polycation/polyanion shell by polymerization. The average size of the encapsulated particles was in the 100–200 nm range. A similar method was adopted by Lvov [85] for producing aqueous nanocolloids of encapsulated drug particles with 150–200 nm diameter. It should be noted that nanoscale synthetic polymer in solution has been widely used in biomedical applications. The more frequently used terminology in biotechnology literature is "nanocolloid", but the concept is essentially same as nanofluids.

Microwave-Assisted Reaction

The use of microwave irradiation has been adopted in various studies for promoting nanoparticle formation in liquids via chemical reactions. Such methods were found to be fast and efficient for preparing nanofluids in one-step synthesis protocols compared to other chemical routes. Zhu [86] prepared copper nanofluids by reducing $CuSO_4 \cdot 5H_2O$ with $NaH_2PO_2 \cdot H_2O$ in ethylene glycol under microwave irradiation. Most of the Cu particles are about 10 nm in diameter, and the nanofluid was found to be stable for more than 2 weeks in the quiescent state at 120 °C. Nikkam [87] fabricated Cu nanofluids in diethylene glycol (DEG) base liquid, by heating up $Cu(Ac)_2 \cdot H_2O$–DEG mixture in microwave oven with PVP as a stabilizer. The average nanoparticle size produced was 75 ± 25 nm. Singh [88] prepared stable silver nanofluid in ethanol by reduction of $AgNO_3$ with polyvinylpyrrolidone (PVP), used as the stabilizing agent, using microwave radiation. Ag nanoparticles with size ranging from 30 to 60 nm were produced with different salt-to-PVP ratio and microwave irradiation duration. Habibzadeh [89] prepared SnO_2 nanofluid in water by the microwave-induced chloride solution combustion synthesis (CSCS) method, in which $SnCl_4$, sorbitol, and ammonium nitrate were heated up to combustion temperature by microwave irradiation. The average particle size was 69 and 153 nm for two different samples. Jalal [90] synthesized zinc oxide nanoparticles in ionic liquid 1-butyl-3-methylimidazolium bis(trifluoromethylsulfonyl) imide, [bmim][NTf_2], by microwave decomposition of zinc acetate precursor.

Microwave irradiation can provide a rapid synthesis technique by uniform heating of reagents and solvents which helps to accelerate the chemical reaction of the metal precursors as well as the nucleation of nanoparticles in the solution. Such features result in monodispersed nanostructures which is beneficial for nanofluid synthesis.

Other Methods

Researchers have also considered various other approaches for generating nanoparticles in base fluids. Teng [91] developed an oxygen-acetylene flame synthesis system to fabricate nanocarbon-based nanofluids (NCBNF) through a one-step method. In the system, the O_2-C_2H_2 flame was served as a carbon source, and the generated smoke was cooled and condensed by water mist to form NCBNF. Kim [92] developed an one-step electrochemical method for producing water-based stable carbon nano colloid (CNC) without adding any surfactants at the room temperature. Carbon nanoparticles were formed by applying electric power to the graphite anode and stainless steel cathode immersed in a DIW bath. The samples were observed to be stable for 30 days and the average size of the suspended nanoparticles was measured to be ~15 nm. Phase transfer method has been developed for preparing homogeneous and stable graphene oxide colloids. Graphene oxide nanosheets (GONs) were successfully transferred from water to n-octane after modification by oleylamine. Kang [93] prepared high-quality carbon nanotube in water by dipping red-hot (> 800 °C) graphite rods into cool water repeatedly. The multiwall nanotubes synthesized in-situ were found to have an inner diameter between 5 and 10 nm and outer diameter between 30 and 50 nm.

6.3 Nanofluid Stabilization

Stabilization is the most important issue in nanofluid research and applications, as the properties of nanofluids could be drastically affected by the clustering and aggregation of nanoparticles. Clustering of nanoparticles in fluid is a natural and spontaneous process due to the strong Brownian motion of liquid molecules which promote collision between nanoparticles, while the high surface energy of nanoparticles promotes adhesion after collisions [94]. In these clusters, the nanoparticles could be held together by either weak physical bonds (intermolecular forces) which are readily broken apart by external forces, or tight chemical bonds (which are difficult to separate). In the previous case, the process is usually reversible as the size of cluster can be reduced easily by means of ultrasonication or stirring. In the latter case, the clustering process is irreversible as very strong forces (i.e., high energy ball milling) are needed to break down the agglomeration [95]. In many cases, nanoparticle precipitates are "weak" agglomerates which can be easily re-dispersed into fine nano-suspensions. In other cases, stable agglomerated particle clusters could still be dispersed uniformly in the suspensions for long times without sedimentation. Hence, the concept of "stable" nanofluid should be clarified into different applications and scenarios.

6.3.1 Characterization of Stability

Although there is no standardized protocol for quantitatively evaluating the level of stability of particle dispersion in nanofluid/noncolloid, three methods have been generally used by different researchers for characterizing and analyzing the stability of nanofluids.

6.3.1.1 Sedimentation and Centrifugation Methods

Sedimentation method is the simplest and most straightforward method for characterization of nanofluid stability. Certain quantity of nanofluids is dispensed into a specific container and the process of nanoparticle sedimentation in the suspension is observed over time. In most studies, a time frame will be reported as an indicator for the stable duration for the sample nanofluid, during which no or little visual sedimentation of particles can be observed visually [57, 96–111]. Many of these studies have also recorded the particle precipitation process using cameras, while few other studies have tried to measure the particle sedimentation quantitatively by recording the drop in solid–liquid interface height [112], thickness of particle sedimentation [113], nanoparticle densities at different height in the nanofluid [110], etc. One significant drawback of the sedimentation method is that it could be extremely time-consuming as some nanofluids were found to be stable over 12 months [114]. In order to evaluate the nanofluid stability faster, centrifugation methods have also been used by a variety of studies, in which visual investigation of sedimentation of nanofluids was performed using a dispersion analyzer centrifuge. Singh [88] confirmed the stability of silver nanofluid by centrifuging the nanofluid sample at 3000 rpm for 10 h. Mehrali [115] prepared graphene nanoplatelets (nanofluids) in distilled water with different mass concentrations. The author observed little sedimentation on the bottom of test tubes after the samples were centrifuged at 6000 rpm for 5–20 min. Fang [116] prepared deep eutectic solvent-based graphene nanofluids and confirmed the stability by centrifugal process (5000–20,000 rpm) for 30 min.

6.3.1.2 Zeta Potential (Electro Kinetic Potential)

Zeta potential is the electrostatic potential between bulk fluid and particle surface induced by the particle surface charge. This indicates the interaction energy between particles, and is in many cases responsible for the stability of particles toward coagulation [117]. Generally speaking, high absolute value of zeta potential means stronger repulsive force between nanoparticles, and hence indicates better stability of nanofluids. Typically, a colloid is considered unstable, moderately stable, stable and highly stable with zeta potential values less than 30 mV, between 30 and 40 mV, between 40 and 60 mV, and greater than 60 mV, respectively [118]. Measurement of zeta potential is a well-developed technique with standardized protocols [119]. The process itself is fast and easy through a variety of commercially available equipment. Hence, this is widely used for characterizing the stability of nanofluids under different conditions [53, 66, 97, 104, 106, 120–122].

6.3.1.3 Spectral Absorbency Analysis

In typical spectral absorbency analysis, the intensity of radiation absorption passing through a target sample is measured for different frequencies. It is possible to characterize the particle size distribution of nanoparticles in nanofluid using absorbance spectrum, since the optical properties of small nanoparticles depend on their morphology (i.e., size and shape). UV–Visible spectroscopy has been adapted for use as a simple and reliable method for monitoring the stability of various nanoparticle solutions including gold [123], copper [124], Al_2O_3 [124–126], ZnO [127], CuO [128], SiO_2 [128], TiO_2 [89, 125, 129], CNT [130], etc. The analytical prediction of the particle size from the spectrum could be achieved using the well-known theory of Mie by Kreibig and Genzel [131]. In general, as the particles become less stable (agglomerate or precipitate), the original extinction peak will decrease in intensity due to the depletion of the fine stable nanoparticles, and often the peak will broaden or a secondary peak will form at longer wavelengths due to the formation of aggregates.

6.3.1.4 Electron Microscopy and Dynamic Light Scattering

One straightforward approach for monitoring the nanofluid stability is to measure the particle size at different time intervals. Scanning/transmitting electron microscope (SEM/TEM) can capture fine image of nanoparticles with resolution down to nanometer scale. The distribution of particle size and evolution of particle coagulation can be directly visualized by observing the particle images. Dynamic light scattering is another commonly used technique for determining particle size in suspension/solution. The process is fast and easy and does not require separation of nanoparticle from the solvent.

6.3.2 Approaches for Enhancing Stability

6.3.2.1 Mechanical Mixing

Mechanical mixing has been widely used in nanofluid preparation for attaining better stability. It is a fast and efficient way for breaking down agglomerated clusters and keeping individual particles separated from each other.

Ultrasonication

Ultrasonication is one of the most commonly used mechanical mixing methods for improving the dispersion stability of nanofluids, in which the longitude sound wave travels through the nanofluid suspension and induces strong oscillations of molecules in the system. Such agitated motion promotes distortion of nano-agglomerates which breaks them into finer particles. Depending on the type of ultrasonicator used, the sonication wave can be applied by either direct or indirect means. In the direct sonication case, a sonication probe is immersed directly into the suspension and the sonication energy is released into the liquid without physical barriers. In the indirect sonication case, the colloidal mixture is usually contained in a vessel which is partially immersed in a sonication bath. The sonication wave was generated on the surface of the bath or chamber, which then travelled through the bath liquid and passed through the container wall before it finally reached the suspension sample. The detailed procedure for performing ultrasonic dispersion of nanofluids was discussed in an NIST protocol by Taurozzi and Hackley [132], in which they have recommended direct sonication over indirect sonication due to the higher effective energy output. However, there is not enough experimental data in support of such claims, as considerable number of reports have used sonication bath for preparing nanofluids which showed good stability over time. Nevertheless, since appreciable amount of sonication energy was absorbed by the suspension container, it generally requires more time and power for bath-sonicator (indirect) for achieving good dispersion compared to the probe-sonicator (direct). As significant amount of heat is released in the ultrasonication process, a shorter sonication period could potentially lower the risk of nanofluid degradation due to overheating, unless a cooling system is incorporated in the ultrasonicator. Generally speaking, the maximum duration of probe sonication is in a time scale of few minutes, while the maximum duration of bath sonication can be extended to more than 24 h.

As stated in the NIST protocol [132], sonication is a highly system-specific dispersion procedure, and suggested that the optimum parameter for sonication power and time varies from sample to sample. The determination of the optimum sonication parameters is a trial-and-error process, in which the researcher should start by referring to literature studies of similar particle–liquid combinations. However, it should be kept in mind that even with the same type of material and particle concentration, the optimum ultrasonication time could be different due to the difference in experimental conditions.

High-Pressure Homogenizer (HPH)

HPH is a powerful and effective method to produce homogenous particle dispersion in liquid, by forcing the nanofluid flowing through a narrow valve under high-pressure conditions. Typical processing pressure of HPH ranges between 20 and 100 MPa, in which the high shear stress ruptures large agglomerates into small and fine particles. The effect of pressure on nanofluid stability has not been studied thoroughly, but it has been shown in few studies that nanofluid prepared via HPH exhibits smaller particle size and hence better stability compared to ultrasonication and other mechanical approach [133, 134]. Hwang et al. [133] examined the TEM images of carbon black (CB) nanoparticles in water-based nanofluid stabilized using stirrer, ultrasonic bath, ultrasonic disruptor, and HPH. It was found that only HPH method can provide sufficient energy to break the particle clusters. Fedele et al. [134] examined average particle size of three different nanofluid samples (CuO/TiO_2/SWCNH - H_2O) dispersed via ball milling, ultrasonication and HPH at 4 and 15 days after synthesis. It was observed that nanofluid stabilized with HPH has the lowest level of agglomeration. However, it is worth noting that very few studies have used HPH for synthesizing nanofluids in which the stability enhancement was only verified in small time scales (less than a month). It is not clear if PHP is really more effective in preventing nanoparticle agglomeration in longer term compared to that of ultrasonication approaches.

6.3.2.2 Dispersant

The addition of dispersants (also referred to as surfactants) is an easy and economic approach for enhancing the stability of nanofluids. The dispersing agents are usually amphiphilic organic molecules with both hydrophobic tail and hydrophilic head group. These dispersants will attach to the surface of the nanoparticle due to the mutual affinity, which helps increase the contact at the interface between the solid particle and base fluid. In addition, the tail of the attached dispersant works as a steric barrier which prevents the particles from agglomerating. Such effect is known as steric hindrance and inhibits the coagulation of nanoparticles in the suspensions. The absorbed layer also enhances zeta potential and promotes electrostatic stabilization effect [135].

The selection of suitable dispersant is dependent on the particle and base fluid. It is suggested that water-soluble surfactants should be used if the base fluid is the polar solvent, while oil-soluble surfactants should be used if the base fluid is non-polar [136]. Such characteristics are usually represented by the hydrophilic–lipophilic balance (HLB) value of the surfactant, which describes the balance of the size and strength of the hydrophilic and hydrophobic groups of the surfactant. It has been reported that surfactants with HLB values greater than 10 have the higher affinity in aqueous solutions, while those with HLB values less than 10 are more oil soluble [137]. One example of oil soluble surfactant is oleic acid, which has extremely low HLB value of one and has been extensively used in the preparation of non-polar nanofluids (i.e., transformer oil, silicone oil, kerosene, etc.). The typical example of water-soluble surfactants is sodium dodecyl sulfate (sodium lauryl sulfate, SDS) with a high HLB value of 40. It has been used in the preparation of various kinds of water-based nanofluids.

Still, it should be noted that the choice of the surfactant does not necessarily have to be consistent with the HLB-based principle. Parametthanuwat et al. [138] prepared aqueous silver nanofluid using oleic acid as dispersing surfactant, which showed effective enhancement of the nanofluid dispersion. Also, most surfactants are organic chemicals which easily degrade at elevated temperatures. Hence, the use of surfactants should be performed in accordance with the actual experimental requirements.

6.3.2.3 pH Control

Adjusting pH value can significantly improve the stability of the nanofluids by changing the zeta potential of the system. By moving away from the isoelectric point (IEP), the surface charge of nanoparticles increases due to the more frequent attachment of surface hydroxyl group. The highly ionic charged state effectively keeps particles apart and hinders agglomeration due to the mutual electrostatic repulsive forces. In practice, such adjustment is usually performed with careful addition of acidic or alkaline chemicals (i.e., HCl or NaOH). For each different nanofluid mixture, there exists an optimum pH condition, at which the absolute value of zeta potential is maximized to ensure that the most stable conditions for the nanofluids are attained. For example, the maximum zeta potential for Cu-H_2O, Al_2O_3-H_2O, and SiC-H_2O nanofluid systems is attained at pH = 8.5~9 [139, 140], pH = 8 [9], and pH = 10 [141], respectively. Such values are dependent on the nanoparticle-fluid system and should be obtained experimentally from parametric studies.

6.3.2.4 Surface Modification

Nanoparticle surfaces can be modified to include different functional groups which enable them to be dispersed in the base fluid. These functional groups could effectively enhance the wettability of the solid–liquid interface, reduce the surface energy of the target nanoparticle and van der Waals forces between particles. Wet chemistry method is the most commonly used approach for performing surface functionalization, in which the target nanoparticle is soaked in the reactant for long enough time at desired temperatures. For example, silica nanoparticles can be grafted with silane groups directly by mixing with trimethoxysilane and stirring vigorously for 48 h at 70 °C [142]. CNTs can be functionalized with carboxyl and hydroxyl groups by submerging in sulfuric/nitric acid solution [143]. These functionalized nanoparticles showed superior dispersion in nanofluids compared to that of their pristine forms. Plasma treatment is another surface functionalization approach, in which the target nanoparticles are functionalized by exposing to a continuous wave of discharged plasma of the desired gas. The technique has been used to coat copper and CNT surfaces with various polar groups which help to enhance their dispersion in water [144, 145].

6.4 Summary

Nanofluids have received tremendous attention in the last 2~3 decades due to their superior properties. They have been reported as promising engineering materials to be used in various applications and industries. However, the exploration of nanofluids has still been mainly limited in laboratory-scale studies. Regardless of the increasing number of research papers and patents relating to nanofluids, the commercialization of nanofluids in the real industrial world has not been successful as the research community anticipated [146]. There are many challenges for application of nanofluids in industries, among which the two most critical issues are the long-term stability of nano-suspensions and the prohibitive cost of nanofluid manufacturing in large quantities. These issues are closely linked with the synthesis process of nanofluids, as it has been shown that the stability of nanofluids is greatly dependent on the preparation procedure. The cost issue of nanofluids is significant in real world application, since most nanoparticles are very expensive comparing to the base fluid. Even at low concentrations (i.e., 1.0%), the material cost of nanofluids could be four or five times that of the base fluid. Consequently, in-situ synthesis of nanofluids from cheap materials/precursors is crucial for practical applications. Despite extensive research on nanofluids, a standard set of procedures for nanofluid synthesis should be developed with proven stability of both morphology and material properties. Due to the complex nature of transport mechanisms in nanofluids, the accomplishment of such goals would require a comprehensive and systematic study for better comprehension and verification of the contradictory research results.

Appendix: Nanofluid Two-Step Synthesis

6.4.1 Metallic Nanoparticle-Based Nanofluids

Metallic nanoparticle-based nanofluids have drawn much interest due to the high thermal conductivity of metals. It is expected that the metallic suspensions can enhance the thermal transport properties of conventional heat transfer fluids which makes them suitable for heat exchanger applications.

Copper Nanofluids

The use of copper nanoparticle in nanofluids is appealing since copper has relatively high resistance to corrosion. Xuan and Li [96] prepared both water-Cu and oil-Cu nanofluids using the two-step method in which copper nanoparticles of about 100 nm diameter are directly mixed in water and mineral oil. For water-based nanofluid, laurate salt at 9% concentration was used as a dispersant which was observed to hold the water-Cu suspension stable for 30 h in a stationary state with minimal amounts of clustering. For oil-based nanofluids, oleic acid at 22% concentration was dispersed in the nanofluid to stabilize the suspension followed by ultrasonication for 10 h. The oil-Cu nanofluid was proved to be stable for 1 week with no sedimentation. Li [97] prepared aqueous copper nanofluid by mixing copper nanoparticle (~25 nm) in purified water at 0.1% concentration (with and without dispersant under different pH conditions). He observed that the addition of CTAB dispersant enhanced the stability of water-Cu suspensions by reducing the diameter of copper nanoparticles from 5560 to 130 nm. The nanofluid sample without dispersants quickly agglomerated, while the sample with dispersant remained stable with no sedimentation after 1 week. The water-Cu nanofluid samples showed maximum zeta potential value at pH = 9.5 which indicates that the suspension exhibits better dispersion in slightly basic environments. Saterlie [147] prepared water-Cu nanofluid by first synthesizing copper nanoparticles (~100 nm) in-situ using chemical reduction method and then re-dispersing them in water. The nanofluid was stabilized by adding SDBS as dispersing agent and ultrasonication for 50 min. He found that by increasing the Cu particle loading from 0.55 to 1.0%, several agglomerates are formed in the nanofluid with nanoparticle size increasing from 120 to 800 nm.

Surfactant-free copper nanofluids were also explored in various studies. Lu [148] prepared surfactant-free water-Cu and ethanol-Cu nanofluids by mixing copper nanoparticles (~20 nm) with the base fluid at 0.2–2% concentration and ultrasonicating for 10 h. The nanofluids were tested in a flat capillary pumped loop and sedimentation of nanoparticles observed on the heated surface. However, the morphology of the working nanofluid was not examined after the test. Kole [98] dispersed copper nanoparticles (~40 nm) in distilled water at 0.5% concentration with ultrasonication for 10 h. The suspensions remained stable for more than 15 days with no significant sedimentation. Garg [57] prepared EG-Cu nanofluids by synthesizing copper nanoparticles using a chemical reduction method, with water as the solvent, and then dispersing them in ethylene glycol using a sonicator. The particle loading is 2.0% and no sedimentation was observed after a few days.

It is difficult to compare these different studies and draw a general conclusion on the effect of dispersing agent and ultrasonication toward the stability of nanofluid samples, since the nanofluid samples are synthesized under different conditions with different material characteristics. The stability of nanofluids is very sensitive to the variation in the size of the nanoparticles, concentration, pH, ultrasonication time, surfactant, etc. It is also important to notice that most studies have only shown stable nanofluids for the limited period (from few hours to 1 week). That suggests preparing stable samples of copper nanofluids via two-step methods to be used in long-term duty-cycles is still a challenging research topic.

Gold and Silver Nanofluids

Gold and silver nanoparticles have also been used in many studies due to their unique optical, electrical, and thermal properties (i.e., high electrical conductivity, stability, and low sintering temperatures). Such properties make them desirable in wide range of applications including diagnostics, antibacterial agents, heat transfer fluids, and optical fluids [149].

Preparation of gold nanofluids via two-step method is rare as most gold nanofluids were synthesized directly in the target base fluid from chemical reduction approach [150–154]. Silver nanofluids can be prepared by mixing manufactured silver nanoparticles in the base fluid. DisKang [155] prepared water/EG-Ag nanofluids at 0.1–0.4% volume fraction by dispersing silver nanoparticles (8–15 nm) into fluids without additives or stabilizers and ultrasonicating for 3 h. The nanofluid was generally stable for 1 day. Oliveira [156] prepared stable water-Ag nanofluid with 0.15% volume loading and 80 nm diameter nanoparticles using high-pressure homogenizer. The stabilization was achieved by placing the mixture in the high-pressure homogenizer and circulating for 30 min at 400–500 bars. The nanofluids were visually verified to be stable for at least 6 months. Hwang [133] prepared silicone oil-Ag nanofluid by dispersing produced silver nanoparticles in the base fluid with the assistance of various physical treatment techniques. With primary particle size of 35 nm and particle weight loading of 0.5%, he found that without any treatment, Ag nanoparticles were highly agglomerated in the pure fluid with an average nanoparticle size of 335 nm. Such values were reduced drastically to 182 nm, 147 nm, 66 nm, and 45 nm, after using stirrer, an ultrasonic bath, an ultrasonic disruptor, and a high-pressure homogenizer, respectively. Warrier [157] prepared water-Ag nanofluid at the concentration of 1 and 2% with nanoparticle size of 20, 30, 50, and 80 nm, respectively. The suspension was stabilized by both polyvinylpyrrolidone (PVP) at a concentration of 0.3% and ultrasonication process (while no settlement was observed for 2 h) after synthesis. Paramethanuwat [158] prepared water-Ag nanofluid at a concentration of 0.5% by repeated magnetic stirring and ultrasonicating after the addition of oleic acid (OA) (at concentration of 0.5, 1, and 1.5%) and potassium oleate surfactant (OAK^+). It was found that the OAK^+ exhibited good adsorption on the silver nanoparticles which helped improve the colloidal stability and non-precipitation period of the silver nanoparticles for up to 48 h.

Gold and silver nanofluids are typically synthesized via one-step method due to their inherent simplicity and competitive costs. It is worth noticing that these nanofluids were shown to be stable over several months—just by physical treatment.

Aluminum Nanofluids

Aluminum nanoparticles are of great interest in a variety of fields due to their high enthalpy of combustion and rapid kinetics. These characteristics make them favorable in fuel engineering field including alloy powder metallurgy parts for automobiles and aircrafts, rocket fuel, igniter, smokes, and tracers [159]. The study of aluminum nanofluid is limited as they are easily oxidized into alumina. Boopathy [160] prepared aluminum nanoparticles (~150 nm) by mechanical milling and dispersing them in distilled water and engine oil with 0.025% volume loading. The nanofluids were stabilized using 1% sodium lauryl (Dodecyl) sulfate as dispersant followed by 10 min of ultrasonication at 20 kHz and for 30 min of magnetic stirring at 1500 rpm. Teipel [161] prepared aluminum nanofluid using paraffin oil and HTPB as the base fluid. The Al particles (~80 nm) were dispersed by stirring for several hours and using ultrasound homogenizer before the suspension was tested for rheological measurements.

Aluminum nanopowders can react with water at high temperature (400–600 °C) to generate hydrogen which enhances fuel combustion [162]. Such feature promotes studies on aluminum nanofluid used in combustion and fuel. Kao [163] prepared aqueous aluminum nanofluid for diesel fuel combustion by producing emulsified nano-aluminum (40–60 nm) liquid using both ultrasonic vibrator and agitator. The work did not discuss the stability of the aluminum suspension. Gan [99] prepared aluminum nanofluid in *n*-decane and ethanol fuels with 80 nm Al nanoparticles at 10% mass loading by stirring the mixture vigorously and ultrasonication in an ice bath for 5 min. He observed that the suspension of *n*-decane/nano-Al remains stable for 10 min while ethanol/nano-Al can last for 24 h without obvious sedimentation. Xiu-tian-feng [164] synthesized stable jet fuel-Al nanofluids with 1.0% mass loading by modifying the surface of aluminum nanoparticles with various chemicals. It was found that oleic acid is the most effective surface modification agent which keeps the suspension stable for more than 2 weeks.

The use of aluminum nanoparticles in fuels and combustion requires high enough concentration for achieving considerable contribution to the energy content. Thus, the stabilization of the nano-mixture suspensions for long enough time is crucial for them to be utilized in liquid fuels.

Iron Nanofluids

Iron nanomaterials are of great interest as iron is among the most useful magnetic materials as well as the most abundant and widely used elements on earth. Doping iron nanoparticles in fluid manifest both fluid and magnetic properties which open new area of applications in electronic device, spacecraft propulsion, material science, biomedical instruments, and so on [165, 166]. Hong and Yang [167] prepared iron nanofluids with ethylene in which the Fe nanocrystalline powder (~10 nm) was first synthesized by chemical vapor condensation process and then re-dispersed in the base fluid with 0.2, 0.3, and 0.4% volume loading using ultrasonication (20 kHz). They observed an increment in thermal conductivity of the nanofluid with increasing sonication time from 10 to 70 min and ascribed it to the improved stability of suspension with prolonged sonication. Sinha [168] prepared EG-iron nanofluid by synthesizing iron nanopowders from chemical reduction method and mixing them in the base fluid under 50 min ultrasonic irradiation in concentrations of 1.0 vol.%. Agglomeration of nanoparticles was observed since the nano-crystallite sizes of the powders were below 20 nm while the average particle size in the fluid was around 500 nm. Xuan and Li [169] prepared magnetic iron nanofluid by directly dispersing Fe nanoparticles (~26 nm) into deionized water with the volume percentage of range from 1.0 to 5.0%. The suspension was stabilized using 1.0–6.0 vol.% sodium dodecylbenzenesulfonate as activator and the nanosamples showed good stability from few hours to 1 week. Gan [170] studied the combustion of iron nanofluid fuels prepared from dispersing iron nanoparticles (25 nm) in *n*-decane/ethanol with 5–20 wt.% concentration by hand mixing and ultrasonication. The nanofluid was stabilized with 0.5 wt.% sorbitan oleate as surfactant and mixture remains suspended for few hours.

Although pure iron exhibits better saturation magnetization, they are highly toxic and very sensitive to oxidation without appropriate surface treatment. In contrast, iron oxide nanoparticles are less sensitive to oxidation and, therefore, can give a better and stable performance [171].

Other Metal Nanofluids

Quite few other metal nanoparticles were also explored for synthesizing energy-efficient nano-suspensions. Naphon [172] prepared titanium nanofluid by mixing 21 nm Ti nanoparticles in de-ionized water and alcohol using an ultrasonic homogenizer. The nanofluid was tested in heat pipe without characterization on the suspension stability. Chopkar [173] prepared Al_2Cu and Ag_2Al dispersed nanofluids using the two-step method, in which Al_2Cu (20–30 nm) and Ag_2Al (30–40 nm) nanoparticles were first produced by mechanical alloying using a high energy planetary ball mill followed by dispersing these particles into ethylene glycol and water with volume fractions from 1.0 to 2.0%. The suspension was homogenized by intensive ultrasonic vibration and magnetic stirring with the addition of 1.0 vol.% oleic acid as surfactant. The sample showed good stability with some tendency of agglomeration during the test. However, metal nanoparticles were found more likely to be oxidized at even low temperature [12] which brings instability in the hydrothermal performance. Besides, metal nanofluids suffer from the issue of quick sedimentation and fouling which makes it challenging to use them in practical applications [174, 175].

Oxide Nanoparticle-Based Nanofluids

Oxide nanomaterials have been intensively used in modern nanotechnology. Their unique physicochemical properties have opened up applications in nanoelectronics, sensors, optics, catalysts, biomedicine, etc. [176–180]. Preparation of nanofluids using oxide nanoparticles via two-step methods is discussed below.

Copper Oxide Nanofluids

Copper oxide nanoparticles are used in diverse applications with a range of useful properties such as high electric/thermal conductivity, electron correlation effect, high atom efficiency, etc. [181, 182]. Choi and Eastman [183] first studied copper oxide nanofluid in water and ethylene glycol, in which they dispersed CuO (~20 nm) nanoparticles produced by gas condensation in the base fluid directly by shaking thoroughly. It was observed that Cu nanoparticles agglomerated into large particles (~100 nm) which could still form the stable suspension in the liquid. Kwak [184] prepared copper oxide nanofluid in ethylene glycol using 10–30 nm Cu nanoparticles at 0.001–1% volume fraction dispersed by ultrasonication. It was found that sonication for 9 h gives the best dispersion and the suspension was stable for 100 days. Namburu [185] prepared copper oxide nanofluid in EG–water mixture (60:40) with a particle size of 29 nm and volume concentration increasing from 1 to 6%. The nanofluid mixture was stirred and agitated thoroughly for 30 min with an ultrasonic agitator for ensuring uniform dispersion. Kulkarni [186] did a similar study by mixing CuO nanoparticles (~29 nm) in deionized water with 5–15% volume fraction. The uniform mixture of nanoparticles in water was attained by thorough stirring and ultrasonicate agitation for half an hour. Li and Peterson [100] prepared water-CuO nanofluid with 29 nm diameter and 2–10% volume fraction. The powder and base fluid were blended by immersing in an ultrasonic bath for 3 h and the suspensions were found to be very stable, with essentially no sedimentation over 7 days. Karthikeyan [101] prepared water/EG-CuO nanofluid using monodispersed CuO particles of 8 nm diameter. The suspension was homogenized by using an ultrasonic horn for 30 min without the addition of surfactant. The study found that the nano-mixture remained stable for more than 3 weeks with particle volume concentration below 1%. Above 1% volume concentration, sedimentation in CuO nanofluid was observed. Rashin [102] prepared copper oxide-coconut oil nanofluid using 20 nm CuO nanoparticles with 0.5–2.5% mass concentration. The nanoparticles were dispersed by only 1 h ultrasonication and the suspension remains stable for 7 days after which the sedimentation starts. Kole [103] prepared stable nanofluid by dispersing 40 nm spherical CuO nanoparticles in gear oil with volume fraction ranging between 0.5 and 2.5%. The mixture was stabilized by mixing with oleic acid, intensive ultrasonication for 4 h and magnetic agitation for 2 h. Although aggregation of CuO nanoparticles was identified with average cluster size ~7 times of the primary diameter, the suspension was stable for more than 30 days without visual sedimentation. Sahooli [120] studied the effect of pH and surfactant concentration on the stability of CuO nanoparticles (4 nm, 1.0 wt.%) in the water-based nanofluid. He proposed that the suspension zeta potential and absorbency were maximized at pH = 8 and 0.095 wt.% PVP, which is the optimum condition for obtaining most stable nanofluid. However, the average particle size measured with PVP surfactant was 63 nm indicating clustering of the nanoparticle.

In general, it was found that the copper oxide nanofluids can be quite stable for moderate time period without the presence of the surfactant, if the nanoparticle concentration is low. The time scale for non-sedimentation could vary from few days to months depending on the ultrasonication and stirring condition. Aggregation of ultrafine nanoparticle is inevitable in CuO nanofluid but the nanoscale cluster can still be stable in the mixture suspension.

Alumina (Al_2O_3) Nanofluids

Alumina nanoparticles are among the most widely used due to their abundance and low cost of mass production. The use of alumina nanoparticles in different base fluids has drawn considerable interest in applications including electronic cooling, deep drilling, thermal energy storage, etc. [187]. Beck [188] dispersed 20 nm diameter alumina nanoparticles in ethylene glycol with mass fraction ranging from 3.26 to 12.2%. The nanofluid was stabilized by ultrasonic mixing for several minutes and the suspension remained uniform during the experiments. Timofeeva [189] prepared nanofluids of alumina particles in water and ethylene glycol with three different particle size (11, 20, and 40 nm) and two different volume concentration (0.01% and 0.1%). The mixture was sonicated continuously for 5–20 h in an ultrasonic bath and highly agglomerated nanoparticles were observed in the experiments. It was found that particle with smaller diameter tends to form larger agglomerates and the agglomeration size increases with time as the sample ages. However, nanosamples were still found to be stable in both water and ethylene glycol. Esmaeilzadeh [190] prepared water-alumina nanofluid using 15 nm Al_2O_3 nanoparticle with 0.5 and 1% volume fractions. The mixture was stabilized through a 4 h process of ultrasonication and electromagnetic stirring. No sedimentation was observed throughout the testing period. Sarathi [191] dispersed 50 nm Al_2O_3 nanoparticles in distilled water by magnetic stirring for 3 h and ultrasonication for few hours. Sedimentation of particles was still observed after the sonication and stirrer was used during the experiment to minimize the sedimentation.

Use of surfactant and pH control can significantly enhance the stability of alumina nanofluid. Sharma [192] prepared stable water-Al_2O_3 nanofluid by mixing SDBS with one-tenth the mass of the nanoparticle (~47 nm) in the suspension. The nanofluid was observed to be stable for over a week if the volume concentration is less than 3%. With higher concentration, some sedimentation was observed. Teng [193] prepared water-Al_2O_3 fluid using 0.3 wt.% chitosan as the dispersing agent. The mixture with 0.5, 1.0, and 3.0 wt.% Al_2O_3 nanoparticles showed good suspension for 1 month during which the sample was placed statically. Jung [121] prepared water-based alumina nanofluid using a horn-type ultrasonic disrupter for 2 h. The nano-suspensions with 0–0.1 vol% Al_2O_3 nanoparticles (45 nm) were observed to be stable for more than 1 month with/without the addition of polyvinyl alcohol surfactant. Khairul made a more comprehensive study on the effects of surfactant toward the stability of Al_2O_3 nanofluid. He used different weight fractions from 0.05 to 0.2% of the dispersant SDBS to stabilize the water-Al_2O_3 (10 nm) nanofluid with nanoparticle weight ratio in the range of 0.05–0.15%. It was found that aggregation of nanoparticle still presents in the fluid, but 0.1% SDBS gives lowest mean aggregation size and maximum zeta potential which is an indication of good stability. Ho [194] prepared 0.1–4 vol.% water-Al_2O_3 nanofluid with the particle size of 33 nm. The nano-suspension was stable for at least 2 weeks after magnetic stirring for 2 h. and adjusting pH value to 3. Jacob [195] used similar method prepare stable suspensions of Al_2O_3 nanoparticles (~50 nm) in de-ionized water with 0.25, 0.5, and 1% volume. The mixture was stabilized by adjusting the pH value away from the isoelectric point and sonication for 5–6 h.

In general, it was found that the alumina nanofluids only exhibit short-term stability with mechanical stabilization methods. The stabilization period was enhanced to months if appropriate amount of dispersant was used.

Titanium Dioxide Nanofluids

Titanium dioxide is being widely used in various consumer goods and products including cosmetics, paints, dyes, plastics, drugs, etc. Nanoscale TiO_2 has high diffraction index and strong light scattering capability which makes them highly used in radiation protection productions, photocatalyst and photovoltaic [196–200]. The use of TiO_2 nanoparticle in nanofluid is promising due to its excellent chemical/physical stability and low cost from commercial manufacturers.

Ding and Wen [201] dispersed 30–40 nm TiO_2 nanoparticles in distilled water with 0.024% volume concentration. The stabilization of nanoparticles in water was realized by (1) cleaning of the bottles in ultrasonic bath; (2) adjusting the pH of the base liquid to pH = 3; (3) ultrasonification of the bottles containing dispersion for 15 min; and (4) shear mixing of the dispersion under the homogenizer for 30–180 min. The dispersion was found to be very stable for at least a couple of weeks without visually observable sedimentation. Murshed [202] prepared titanium dioxide nanofluid by dispersing TiO_2 nanoparticles in rod-shapes of ∅10 nm × 40 nm (diameter by length) and in spherical shapes of ∅15 nm in deionized water. The nanoparticles were dispersed uniformly using ultrasonic dismembrator for 8–10 h with/without 0.01–0.02 vol.% oleic acid and CTAB surfactants. It was found that nanoparticles agglomerated into large clusters without surfactant, and adding surfactant brought better stability which is indicated by the increment in nanofluid thermal conductivity. Turgut [203] prepared water-based TiO_2 nanofluid with particle size of 21 nm and particle volume concentration from 0.2 to 3%. The mixture was homogenized using ultrasonic vibration which breaks down the agglomerations. Yue-fan [204] prepared titanium dioxide colloidal suspension by dispersing TiO_2 nanoparticles (<20 nm) in transformer mineral oil with 0.003–0.05 g/L concentration. The particles were dispersed by just ultrasonic route and the suspension was stable for 24 h. He and Jin [205] prepared aqueous TiO_2 nanofluid by dispersing 20 nm diameter dry titanium dioxide nanoparticles in distilled water with 1.0, 2.5, and 4.9% mass concentration. The mixture was stabilized by first applying ultrasonication for 30 min, then processed in a medium-mill and finally adjusting the pH value to 11. The particle size distribution after each process stage was ~500 nm, ~120 nm, and ~95 nm while the nanofluids were found to be very stable for months. Charkraborty [206] prepared 0.1 wt.% water-based TiO_2 nanofluid using dry particles with size in the range of 30–50 nm. The nanofluid was homogenized by high shear mixer which breaks down the agglomerate and addition of 0.01 wt.% surfactant which ensures longer stability. Fedele [207] studied the characterization of water-based nanofluids containing TiO_2 (~72 nm) nanoparticles in mass concentrations ranging between 1 and 35%. The nanofluids were stabilized using 1–5 wt.% acetic acid as dispersant with 1 h sonication. The mean diameter of the static suspension decreases to around 51 nm after 35 days, indicating a partial precipitation. Such value returned to 76 nm after re-sonication for 1 h, suggesting the absence of further aggregation in the suspension. Said [208] dispersed 21 nm TiO_2 spherical particles in distilled water with 0.1% and 0.3% volume concentration. The homogenous dispersed solution was obtained after adding PEG 400 dispersant with two times the mass of the particles and passing through 30 cycles in a high-pressure homogenizer. No visual sign of aggregation and sedimentation was observed for a period of a month. Muthusamy [209] prepared stable titanium dioxide nanofluid by dispersing 50 nm diameter TiO_2 particles in ethylene glycol with 0.5, 1.0, and 1.5% volume concentration. The suspension was stabilized by merely mechanical stirring process and proved to be stable for more than 3 weeks with ~220 nm local clusters.

The titanium dioxide nanofluid exhibits relatively good stability (from few weeks to months). Different stabilization conditions were required for achieving optimum dispersion depending on the base fluid and particle characteristics.

Zinc Oxide Nanofluids

Zinc oxide is also among some of the widest used nanomaterials with its good electrical, electrochemical, and structural properties [210, 211]. ZnO nanoparticles exhibit in various form (particle, rod, thin film) and can be used in electroluminescent devices, chemical sensors, solar cells, etc. [212]. Zhang [213] prepared water-based zinc oxide nanofluid with 20 g/L concentration by dissolving ZnO nanoparticles (60–200 nm) in distilled water, sonicating for 30 min and milling for another 3 h. The average particle size increased slightly from 198.4 to 225.9 nm after being stored for 120 days, indicating good stability of the nanosuspension over time. Yu [214] dispersed dry ZnO nanoparticles (10–20 nm) in ethylene glycol with volume concentration ranging from 0.2 to 5%. The nanofluid was stirred and sonicated continuously for 15 min to 12 h. It was found that the average particle size decreases rapidly in the first 3 h and remained 210 nm afterward. The measured average particle size in the formulated nanofluids is much larger than the size of primary particles indicating ultrasonification was not able to break the agglomerates completely. Sagadevan [215] prepared ZnO nanoparticles (15–20 nm) first by solvothermal reaction and dispersed them in polyvinyl alcohol with magnetic stirring process and ultrasonic vibrator for 5 h. The dispersed mixture was clear and stable for up to 2 weeks. Esfe and Saedodin [216] prepared EG-ZnO nanofluid using 18 nm ZnO nanoparticles with volume concentration ranging from 0.25 to 5.0%. The suspensions were subjected to ultrasonic vibrator for 3–5 h in order to get a uniform dispersion and a stable suspension. Subramaniyan [104] prepared water-ZnO nanofluid by dispersing 0.1%–0.4 wt.% ZnO nanoparticles in water using ultrasonication for 20 min. It was found that ZnO nanofluids with 0.3 wt.% showed highest stability with the maximum zeta potential values. Visual sedimentation showed that 0.3 wt.% nanofluid is stable for 20–24 h without any trace of sedimentation but all the other fluids settle within 6–12 h. Raykar and Singh [105] synthesized water-soluble ZnO nanoparticle (non-spherical, 100–150 nm) via chemical precipitation method and dispersed them in deionized water. The mixture was sonicated for 1 h with the addition of acetylacetone as dispersant. The nanofluid was found to be stable over 9 months to 1 year. Suganthi and Rajan [106] prepared stable ZnO–water nanofluids with particle volume concentrations in the range of 0.25–2 vol.%. They dispersed ZnO nanoparticles (35–45 nm) into water with sodium hexametaphosphate (SHMP) surfactant under high shear homogenization for 20 min, followed by ultrasonication for 180 min. The high colloidal stability was verified by high absolute value of zeta potential as well as visual observation. Saliani [107] dispersed ZnO nanoparticles (4.45 nm) in glycerol with the aid of a magnetic stirrer. Ammonium citrate with the same mass of nanoparticles were used as a dispersant to enhance the stability of the nanofluids, and the suspensions were stable for at least several months with no sedimentation observed during the period.

It is noticed that the stability of zinc oxide nanofluids could be significantly enhanced by adding proper surfactant which could be potentially helpful for using them in the long-term application.

Iron Oxide Nanofluids

Iron oxide nanoparticles have attracted considerable interest due to their superparamagnetic properties and their potential biomedical applications arising from its biocompatibility and non-toxicity [217]. Asadzadeh [108] dispersed 0.05 vol% and 0.1 vol.% Fe_3O_4 nanoparticles (<50 nm) in ethylene glycol using vigorous mechanical agitation and ultrasonication for 1 h. The suspension was stable for 12 h without observable sedimentation. Sheikhababai [109] prepared Fe_3O_4 nanofluid using EG–water mixture (50 vol.%-50 vol.%) with 0.02–0.1% particle volume loading. The 50 nm diameter Fe_3O_4 nanoparticles were added into the base fluid gradually under ultrasonic mixing for an hour. The nanofluid was stabilized with another hour of sonication and no sedimentation was observed for 8 h. Sundar [110] prepared water-based nanofluid using 36 nm Fe_3O_4 nanoparticles at 0.02, 0.1, 0.3, and 0.6% volume concentrations. The particles were uniformly dispersed in the base fluid at pH value of 3 with 2 h sonification. The uniform dispersion of the nanoparticles is established by visual observation for nanoparticle sedimentation and measuring the densities of nanofluid at different locations in the container. Župan and Renjo [111] prepared water-based ferrofluids by dispersing 50 nm diameter Fe_3O_4 nanoparticles in deionized water using ultrasonic bath for 90 min. The sonified colloid was stable for 1 h without dispersant or activating agent. However, visible sedimentation was observed in the bottom of the suspension after 24 h. Phuoc and Massoudi [218] dispersed Fe_2O_3 nanoparticles (20–40 nm) in deionized water with 0.2 wt.% Polyvinylpyrrolidone (PVP) or Poly(ethylene oxide) (PEO) as surfactant. The suspension was homogenized by magnetic mixing and ultrasonication for 30 min. It was observed that these nanofluids could remain stable for 2 weeks if the particle concentration is less than 2% and less than 1 week if the concentration is higher. Goshayeshi [219] prepared γ- and α-Fe_2O_3/Kerosene nanofluids with 2.0% volume concentration. The nanoparticles were added into the base fluids with oleic acid surfactant and stirred constantly, followed by 5 h sonication. The Fe_2O_3 nanoparticles could readily disperse in organic solvent and the suspension was stable for 10 days. Salari [220] prepared aqueous iron oxide nanofluids by dispersing 0.1–0.3 wt.% Fe_3O_4 nanoparticles (~20 nm) into the water using motorized magnetic stirrer with speed of 250 rpm for 30 min. The suspension was stabilized by adding 0.1 vol.% nonylphenol ethoxylate surfactant into DIW, ultrasonication for 30 min and adjusting pH values. The most stable nanofluid was obtained when pH = 8.43 and the suspension was stable for 25 days.

It can be seen from various studies that the iron oxide nanofluids exhibit relatively shorter stabilization period (less than a month) even with pH control and surfactant.

Silicon Dioxide Nanofluids

Silicon dioxide nanoparticles are of great interest in a variety of biomedical applications due to their stability, low toxicity and capability for functionalization with different molecules and polymers [221]. Fazeli [222] prepared water-SiO_2

nanofluids by dispersing 18 nm silica nanoparticles in distilled water with 3.5, 4, 4.5, and 5% volume concentration. The suspension was stabilized using ultrasonic bath for 90 min without any surfactant and the nanofluids were stable for a period of 72 h without any visible settlement. Jin [223] prepared 0.005–0.1% mass fraction silica nanofluid in mineral oil using 10–20 nm size SiO_2 nanoparticle. The particles were dispersed uniformly in the base fluid using magnetic stirring for 15 min and ultrasonication for 2 h. The 0.005% and 0.01% silica nanofluids were found to be stable for around 1 month, for the 0.02% silica nanofluid the stability of the suspension was reduced to 2 days, and the 0.1% silica nanofluid was stable for less than 24 h. Rafati [224] prepared silica nanofluid using a mixture of deionized water (75 vol.%) and ethylene glycol (25 vol.%) as the base fluid. SiO_2 nanoparticles with 14 nm average size were dispersed in the base fluid with 0.5, 1.0, and 1.5% volume concentration using ultrasonication. The nanofluids showed great stability even after 1 week. Noghrehabadi [225] dispersed 12 nm SiO_2 nanoparticles in water with 1% mass concentration using vertical mixer and ultrasonication for 60 min. The nanofluid was homogenized in ultrasonic bath every day to break down the agglomeration and minimize the sedimentation. Sharif [226] prepared polyalkylene glycol-SiO_2 nanolubricant by dispersing 30 nm SiO_2 nanoparticles in the base fluid using a magnetic stirrer for 1 h, and then surged using ultrasonic bath vibrator for 2 h. Minimum sedimentation was observed 1 month after the preparation of nanofluid with 0.2–1.5% volume concentration. Liu and Liao [227] prepared silica nanofluid in both water and alcohol. SiO_2 nanoparticles with 35 nm average diameter were dispersed in the base fluid with 0.2–2% mass concentration. The nanofluids were mixed with 0.5 vol.% SDBS surfactant and surged in super-sonic water bath for 12 h. The experimental results showed that the stability and uniformity of nanofluids were good at least in 1 month. Zhang [122] studied the influence of ultrasonication, dispersants, and pH on the stability of water-silica nanofluids. The nanofluids were prepared by dispersing 1.0 wt.% SiO_2 nanoparticles (~50 nm) in water with mechanical force agitation, ultrasonication, and addition of SDBS. It was found that the silica clusters were effective dispersed with average size of 63 nm in suspension under the sonication power of 500 W and sonication time of 120 min. The maximum absolute zeta potential was attained with SDBS concentration of 1.0% and pH value of 9.5. The good stability was also verified with long-term test in which the particle size remained unchanged after 7 days. Yang and Liu [114] prepared stable water-based nanofluid by dispersing surface-functionalized SiO_2 nanoparticles (30 nm) in water with 10% mass fraction. The nanofluid was kept stable for 12 months. Bagwe [228] prepared silica nanoparticles with different functional groups (including carboxylate, amine, amine/phosphonate, polyethylene glycol, octadecyl, and carboxylate/octadecyl groups) in water and studied their aggregation behavior in water. It was found that the nanoparticles prepared with appropriate amount of amine/phosphonate functional group were stable for more than 8 months in aqueous solution.

The silica-based nanofluids have relatively low stability using only physical stabilization method. The addition of surfactant has some marginal effect on the improvement of the stability, but surface modification could effectively make silica nanoparticles sustain much longer in the suspension environment.

Organic Nanofluids

Organic nanoparticles exhibit superior electric and thermal properties owing to their unique metal lattice or graphite structures, and have attracted attention for a wide range of applications in different fields. Dispersing carbon-based nanoparticles in liquid has not been as easy as other nanoparticle, as carbon-based particles have a strong tendency to agglomerate due to the strong intermolecular force. It has shown that pristine CNTs will precipitate rapidly in most of fluids even with prolonged sonication [229]. Consequently, surfactant and surface modifications have been used in almost all carbon-based nanofluid preparations.

Wen and Ding [230] prepared the stable aqueous suspension of MWCNT with 0–0.84% volume concentration. The prepared sample was stabilized following a sequence of steps involving: (1) ultrasonicating CNT sample in water bath for 36 h; (2) dispersing CNT in distill water with 20% by weight of sodium dodecyl benzene sulfonate (SDBS) with respect of CNTs; (3) ultrasonicating mixture in water bath for 24 h; (4) treating suspension with high-speed magnetic stirrer for 1 h. The aqueous CNT nanofluid was found to be very stable from months without sedimentation. Wusiman et al. [231] prepared MWCNTS in aqueous suspension with surfactant SDBS and sodium dodecyl sulfate (SDS). They changed the MWCNT concentration from 0.1 to 1 wt.%, with CNT/surfactant ratio varied from 1/1 to 4/1. The suspension was only subjected to ultrasonic mixing for 20 min and the samples with 3/1 CNT-surfactant ratio was found to be most stable for more than 1 month without sedimentation. SDBS was found slightly superior than SDS in their study with better thermal performance. Rashmi et al. [232] prepared aqueous dispersion of CNTs in the presence of gum arabic (GA), with concentrations of CNT and GA varying in the range of 0.01–0.1 wt.% and 0.25–5 wt.% respectively. The mixture was homogenized at 28,000 rpm for 10 min and further sonicated in water bath for 1–24 h. It was found that the optimum concentration of GA varies from 1.0 to 2.5 wt.% with increasing CNT concentration, and the nanofluid was found to be stable for more than 40 days. Quite few other surfactants have also been tested to show effective enhancement on the suspension stability of CNT nanofluid including polyvinylpyrrolidone (PVP) [233], hexadecyltrimethyl ammonium bromide (CTAB) [234], chitosan [235], and gemini surfactant [236]. The optimum concentration for each surfactant is dependent on its own property and the interaction with the carbon molecules.

Pre-treating CNTs in acid endows them with carboxylic acid and hydroxyl groups, which could effectively prevent the CNTs from aggregating over time [237]. Osorio et al. prepared functionalized CNTs by soaking CNTs in three different acid environments for 2 h: (1) $H_2SO_4/HNO_3/HCl$; (2) H_2SO_4/HNO_3; and (3) HNO_3. The treated CNTs were dispersed in water and the subsequent sedimentation over 20 days showed the good stability of the sample soaked in multi-acid environment. Zhang et al. [238] prepared water-soluble CNTs by the introduction of potassium carboxylate (–COOK) using potassium persulfate (KPS) as oxidant. The KPS-treated SWNTs was dispersed in deionized water with ultrasonic water bath and found to be stable over 1 month. Narisi et al. [239] prepared stable CNT nanofluid in water using a combination approach of surface functionalization, surfactant, and ultrasonication. The CNTs were first treated with KPS oxidant, and then dispersed in water with 0.25 wt.% SDS undergoing 45 mins' ultrasonication using both probe and bath ultrasonicator. The average particle diameter was examined using dynamic light scattering which remained constant (~200 nm) 2 months after preparation.

Although we have been discussing CNT-based nanofluid here, the concept and preparation method is quite similar for graphene/graphene oxide/fullerene-based nanofluid. Owing to their common surface properties, the dispersing method is widely acceptable between different types of carbon nanoparticles.

Special Nanofluid (High/Low Temperature)

For very high-temperature and low-temperature nanofluid, the preparation of nanofluid involves special steps as the mixing process cannot be performed in room environment. These nanofluids are usually based on materials which are not in liquid form at room temperature and atmospheric pressure (i.e., molten salt, liquified gas, low temperature refrigerant).

High-Temperature Nanofluid

For materials which are in the solid state at room environment, the preparation of nanofluid is usually performed by first dissolving the target material in a room temperature liquid (usually water), then dispersing nanoparticle in the solution, and finally evaporating water out and heating the composite to high temperature where it transformed into liquid. A typical example is the molten salt-based nanofluid which melts at temperature more than 200 °C. Shin and Banerjee [240] prepared silica nanofluids in alkali chloride eutectic. They first dissolve all chloride salt in distilled water, then dispersed 1.0 wt.% SiO_2 nanoparticle via ultrasonication bath for 100 min, and evaporated water in the vial on a hot plate at 200 °C until dried completely. The nanofluid showed good stability as particle size remained constant after repeated DSC cycling tests. Jo and Banerjee [241] prepared graphite nanofluid in molten carbonate salt using gum arabic as the surfactant. In his study, the surfactant and graphite nanoparticle were first dispersed in distilled water with 2 h sonication, then require amount of K_2CO_3 and Li_2CO_3 were dissolved in the suspension liquid with additional 3 h sonication. The final mixture was transferred to a petri dish and heated on a hot plate at 100 °C until fully dehydration. The nanomaterials showed good dispersibility with consistent thermophysical property measurements from repeated tests. Most other molten salt nanofluid preparations [242–246] have followed the same procedure used by Shin and Jo.

Low-Temperature Nanofluid

Most of the widely used refrigerants are in vapor state under room environment (i.e., R134a, R410a). Hence, preparing well-dispersed nanofluid using these materials is usually accomplished by first dispersing nanoparticle in a secondary fluid, and then putting the dispersed fluid into the refrigeration system before the refrigerant fills the test loop. Bi et al. [247] mixed TiO_2 nanoparticle into R134a by first dispersing the nanoparticle into mineral oil via conventional approaches, then put the mixture into the compressor to let the refrigerant mixing with the nanoparticle. Jwo et al. [248] followed a similar approach to mix Al_2O_3 nanoparticles in R134a using POE oil. Subramani and Prakash [249] prepared SUNISO 3GS oil-based nanolubricant with 0.06 wt.% Al_2O_3 nanoparticle which is stable for 3 days without coagulation or deposition. They then filled the nanolubricant in the compressor where it mixes with R134a.

In certain cases, nanoparticles can be added directly into the low temperature liquid as well. Anderson [250] dispersed MWCNT into liquid oxygen (LOX) by tipping the nanoparticles into the LOX gently and slowly. It is mentioned that great cares were taken at this point to avoid micro-scale boiling. The dispersion was then performed by ultrasonicating the mixture with a pre-cooled probe sonicator for 20 s. However, the study on cryogenic nanofluid is rather limited, which may be due to the inherent technical difficulty and narrow application field.

References

1. Chol S (1995) Enhancing thermal conductivity of fluids with nanoparticles. ASME-Publications-Fed 231:99–106
2. Rao CNR, Müller A, Cheetham AK (2006) The chemistry of nanomaterials: synthesis, properties and applications. Wiley, New York
3. Kim S-Y et al (2013) Development of a ReaxFF reactive force field for titanium dioxide/water systems. Langmuir 29(25):7838–7846
4. Miao M et al (2012) Activation of water on the TiO2 (110) surface: the case of Ti adatoms. J Chem Phys 136(6):064703
5. Sun C et al (2010) Titania-water interactions: a review of theoretical studies. J Mater Chem 20(46):10319–10334
6. Mattioli G et al (2008) Short hydrogen bonds at the water/TiO2 (anatase) interface. J Phys Chem C 112(35):13579–13586
7. Kharissova OV, Kharisov BI, de Casas Ortiz EG (2013) Dispersion of carbon nanotubes in water and non-aqueous solvents. RSC Adv 3(47):24812–24852
8. Cardellini A et al (2016) Thermal transport phenomena in nanoparticle suspensions. J Phys Condens Matter 28(48):483003
9. Zhu D et al (2009) Dispersion behavior and thermal conductivity characteristics of Al_2O_3–H_2O nanofluids. Curr Appl Phys 9(1):131–139
10. Pfeiffer C et al (2014) Interaction of colloidal nanoparticles with their local environment: the (ionic) nanoenvironment around nanoparticles is different from bulk and determines the physico-chemical properties of the nanoparticles. J R Soc Interface 11(96):20130931

11. Kraynov A, Müller TE (2011) Concepts for the stabilization of metal nanoparticles in ionic liquids. InTech Open Access, Rijeka
12. Nine MJ et al (2014) Is metal nanofluid reliable as heat carrier? J Hazard Mater 273:183–191
13. Kimoto K et al (1963) An electron microscope study on fine metal particles prepared by evaporation in argon gas at low pressure. Jpn J Appl Phys 2(11):702
14. Akoh H et al (1978) Magnetic properties of ferromagnetic ultrafine particles prepared by vacuum evaporation on running oil substrate. J Cryst Growth 45:495–500
15. Wagener M, Murty B, Guenther B (1996) Preparation of metal nanosuspensions by high-pressure DC-sputtering on running liquids. In: MRS proceedings. Cambridge University Press, Cambridge
16. Eastman JA et al (2001) Anomalously increased effective thermal conductivities of ethylene glycol-based nanofluids containing copper nanoparticles. Appl Phys Lett 78(6):718–720
17. Pike-Biegunski M, Biegunski P, Mazur M (2005) Colloid, method of obtaining colloid or its derivatives and applications thereof. Google Patents
18. Lee G-J et al (2012) Thermal conductivity enhancement of ZnO nanofluid using a one-step physical method. Thermochim Acta 542:24–27
19. Park EJ, Park HW (2011) Synthesis and optimization of metallic nanofluids using electrical explosion of wire in liquids. In: Nanotech 2011 Vol. 2: Electronics, Devices, Fabrication, MEMS, Fluidics and Computational Micro & Nano Fluidics. CRC Press, Boca Raton
20. Kim C-K, Lee G-J, Rhee C-K (2010) Synthesis and characterization of Cu nanofluid prepared by pulsed wire evaporation method. J Kor Powder Metall Inst 17(4):270–275
21. Jafarimoghaddam A et al (2017) Experimental study on Cu/oil nanofluids through concentric annular tube: a correlation. Heat Transf Asian Res 46(3):251–260
22. Khoshvaght-Aliabadi M et al (2016) Performance of agitated serpentine heat exchanger using metallic nanofluids. Chem Eng Res Des 109:53–64
23. Kim C-K, Lee G-J, Rhee C-K (2009) Synthesis and characterization of silver nanofluid using pulsed wire evaporation method in liquid-gas mixture. Kor J Mater Res 19(9):468–472
24. Munkhbayar B et al (2013) Surfactant-free dispersion of silver nanoparticles into MWCNT-aqueous nanofluids prepared by one-step technique and their thermal characteristics. Ceram Int 39(6):6415–6425
25. Park EJ et al (2011) Optimal synthesis and characterization of Ag nanofluids by electrical explosion of wires in liquids. Nanoscale Res Lett 6(1):223
26. Yun G et al (2011) Effect of synthetic temperature on the dispersion stability of gold nanocolloid produced via electrical explosion of wire. J Nanosci Nanotechnol 11(7):6429–6432
27. Kim CK, Lee G-J, Rhee CK (2012) A study on heat transfer characteristics of spherical and fibrous alumina nanofluids. Thermochim Acta 542:33–36
28. Lee G-J et al (2011) Characterization of ethylene glycol based Tio_2 nanofluid prepared by pulsed wire evaporation (PWE) method. Rev Adv Mater Sci 28:126–129
29. Chang H, Chang Y-C (2008) Fabrication of Al_2O_3 nanofluid by a plasma arc nanoparticles synthesis system. J Mater Process Technol 207(1–3):193–199
30. Teng T-P, Cheng C-M, Pai F-Y (2011) Preparation and characterization of carbon nanofluid by a plasma arc nanoparticles synthesis system. Nanoscale Res Lett 6(1):293
31. Tsung T-T et al (2003) Development of pressure control technique of an arc-submerged nanoparticle synthesis system (ASNSS) for copper nanoparticle fabrication. Mater Trans 44(6):1138–1142
32. Chang H et al (2004) TiO2 nanoparticle suspension preparation using ultrasonic vibration-assisted arc-submerged nanoparticle synthesis system (ASNSS). Mater Trans 45(3):806–811
33. Lo C-H, Tsung T-T, Chen L-C (2005) Shape-controlled synthesis of cu-based nanofluid using submerged arc nanoparticle synthesis system (SANSS). J Cryst Growth 277(1):636–642
34. Lo C-H, Tsung T-T, Lin H-M (2007) Preparation of silver nanofluid by the submerged arc nanoparticle synthesis system (SANSS). J Alloys Compd 434:659–662
35. Chih-Hung L, Tsung T-T, Liang-Chia C (2005) Ni nano-magnetic fluid prepared by submerged arc nano synthesis system (SANSS). JSME Int J Ser B Fluids Therm Eng 48(4):750–755

36. Saito G, Akiyama T (2015) Nanomaterial synthesis using plasma generation in liquid. J Nanomater 16(1):299
37. Kazakevich P et al (2006) Laser induced synthesis of nanoparticles in liquids. Appl Surf Sci 252(13):4373–4380
38. Phuoc TX, Soong Y, Chyu MK (2007) Synthesis of Ag-deionized water nanofluids using multi-beam laser ablation in liquids. Opt Lasers Eng 45(12):1099–1106
39. Kim HJ, Bang IC, Onoe J (2009) Characteristic stability of bare au-water nanofluids fabricated by pulsed laser ablation in liquids. Opt Lasers Eng 47(5):532–538
40. Sadrolhosseini AR et al (2016) Green fabrication of copper nanoparticles dispersed in walnut oil using laser ablation technique. J Nanomater 2016:62
41. Tran P, Soong Y (2007) Preparation of nanofluids using laser ablation in liquid technique. National Energy Technology Laboratory (NETL), Pittsburgh and Morgantown
42. Torres-Mendieta R et al (2016) Characterization of tin/ethylene glycol solar nanofluids synthesized by femtosecond laser radiation. ChemPhysChem 18(9):1055–1060
43. Chun S-Y et al (2011) Heat transfer characteristics of Si and SiC nanofluids during a rapid quenching and nanoparticles deposition effects. Int J Heat Mass Transf 54(5):1217–1223
44. Lee SW, Park SD, Bang IC (2012) Critical heat flux for CuO nanofluid fabricated by pulsed laser ablation differentiating deposition characteristics. Int J Heat Mass Transf 55(23):6908–6915
45. Piriyawong V et al (2012) Preparation and characterization of alumina nanoparticles in deionized water using laser ablation technique. J Nanomater 2012:2
46. Thongpool V, Asanithi P, Limsuwan P (2012) Synthesis of carbon particles using laser ablation in ethanol. Procedia Eng 32:1054–1060
47. Choi M-Y et al (2011) Ultrastable aqueous graphite nanofluids prepared by single-step liquid-phase pulsed laser ablation (LP-PLA). Chem Lett 40(7):768–769
48. Zeng H et al (2012) Nanomaterials via laser ablation/irradiation in liquid: a review. Adv Funct Mater 22(7):1333–1353
49. Goya GF (2004) Handling the particle size and distribution of Fe_3O_4 nanoparticles through ball milling. Solid State Commun 130(12):783–787
50. Ghazanfari N et al (2007) Preparation of nano-scale magnetite Fe_3O_4 and its effects on the bulk Bi-2223 Superconductors. In: AIP Conference Proceedings, AIP
51. Can MM et al (2010) Effect of milling time on the synthesis of magnetite nanoparticles by wet milling. Mater Sci Eng B 172(1):72–75
52. Inkyo M et al (2008) Beads mill-assisted synthesis of poly methyl methacrylate (PMMA)-TiO_2 nanoparticle composites. Ind Eng Chem Res 47(8):2597–2604
53. Harjanto S et al (2011) Synthesis of TiO_2 nanofluids by wet mechanochemical process. In: AIP Conference Proceedings, AIP
54. Nine MJ et al (2013) Highly productive synthesis process of well dispersed Cu_2O and cu/Cu_2O nanoparticles and its thermal characterization. Mater Chem Phys 141(2):636–642
55. Almásy L et al (2015) Wet milling versus co-precipitation in magnetite ferrofluid preparation. J Serb Chem Soc 80(3):367–376
56. Liu M-S et al (2006) Enhancement of thermal conductivity with cu for nanofluids using chemical reduction method. Int J Heat Mass Transf 49(17):3028–3033
57. Garg J et al (2008) Enhanced thermal conductivity and viscosity of copper nanoparticles in ethylene glycol nanofluid. J Appl Phys 103(7):074301
58. Kumar SA et al (2009) Synthesis and characterization of copper nanofluid by a novel one-step method. Mater Chem Phys 113(1):57–62
59. Shenoy US, Shetty AN (2014) Simple glucose reduction route for one-step synthesis of copper nanofluids. Appl Nanosci 4(1):47–54
60. Tsai C et al (2004) Effect of structural character of gold nanoparticles in nanofluid on heat pipe thermal performance. Mater Lett 58(9):1461–1465
61. Xun F et al (2005) Preparation of concentrated stable fluids containing silver nanoparticles in nonpolar organic solvent. J Dispers Sci Technol 26(5):575–580

62. Salehi J, Heyhat M, Rajabpour A (2013) Enhancement of thermal conductivity of silver nanofluid synthesized by a one-step method with the effect of polyvinylpyrrolidone on thermal behavior. Appl Phys Lett 102(23):231907
63. Cao H et al (2006) Shape-and size-controlled synthesis of nanometre ZnO from a simple solution route at room temperature. Nanotechnology 17(15):3632
64. Darezereshki E, Ranjbar M, Bakhtiari F (2010) One-step synthesis of maghemite (γ-Fe_2O_3) nano-particles by wet chemical method. J Alloys Compd 502(1):257–260
65. Manimaran R et al (2014) Preparation and characterization of copper oxide nanofluid for heat transfer applications. Appl Nanosci 4(2):163–167
66. Chakraborty S et al (2015) Synthesis of Cu–Al layered double hydroxide nanofluid and characterization of its thermal properties. Appl Clay Sci 107:98–108
67. Abareshi M et al (2010) Fabrication, characterization and measurement of thermal conductivity of Fe_3O_4 nanofluids. J Magn Magn Mater 322(24):3895–3901
68. Zhu H et al (2011) Preparation and thermal conductivity of CuO nanofluid via a wet chemical method. Nanoscale Res Lett 6(1):181
69. Mohammadi M et al (2011) Inhibition of asphaltene precipitation by TiO_2, SiO_2, and ZrO_2 nanofluids. Energy Fuel 25(7):3150–3156
70. Suganthi K, Rajan K (2012) Effect of calcination temperature on the transport properties and colloidal stability of ZnO–water nanofluids. Asian J Sci Res 5:207–217
71. Brinker CJ, Scherer GW (2013) Sol-gel science: the physics and chemistry of sol-gel processing. Academic, New York
72. Birlik I et al (2014) Preparation and characterization of TiO_2 nanofluid by sol-gel method for cutting tools. AKU J Sci Eng 14
73. Leena M, Srinivasan S (2015) Synthesis and ultrasonic investigations of titanium oxide nanofluids. J Mol Liq 206:103–109
74. Gangwar J et al (2014) Strong enhancement in thermal conductivity of ethylene glycol-based nanofluids by amorphous and crystalline Al2O3 nanoparticles. Appl Phys Lett 105(6):063108
75. Subramaniyan A et al (2015) Preparation and stability characterization of copper oxide nanofluid by two step method. In: Materials science forum. Trans Tech
76. Guan BH et al (2014) An evaluation of iron oxide nanofluids in enhanced oil recovery application. In: AIP Conference Proceedings, AIP
77. Lee KC et al (2016) Effect of zinc oxide nanoparticle sizes on viscosity of nanofluid for application in enhanced oil recovery. J Nano Res 38:36–39
78. Sarafraz M, Kiani T, Hormozi F (2016) Critical heat flux and pool boiling heat transfer analysis of synthesized zirconia aqueous nano-fluids. Int Commun Heat Mass Transfer 70:75–83
79. Kim W-G et al (2008) Synthesis of silica nanofluid and application to CO2 absorption. Sep Sci Technol 43(11–12):3036–3055
80. Jing D et al (2015) Preparation of highly dispersed nanofluid and CFD study of its utilization in a concentrating PV/T system. Sol Energy 112:30–40
81. Solans C et al (2005) Nano-emulsions. Curr Opin Colloid Interface Sci 10(3):102–110
82. Han Z, Cao F, Yang B (2008) Synthesis and thermal characterization of phase-changeable indium/polyalphaolefin nanofluids. Appl Phys Lett 92(24):243104
83. Kim G et al (2008) Nanoscale composition of biphasic polymer nanocolloids in aqueous suspension. Microsc Microanal 14(05):459–468
84. Pattekari P et al (2011) Top-down and bottom-up approaches in production of aqueous nanocolloids of low solubility drug paclitaxel. Phys Chem Chem Phys 13(19):9014–9019
85. Ariga K et al (2011) Layer-by-layer self-assembled shells for drug delivery. Adv Drug Deliv Rev 63(9):762–771
86. Zhu H-t, Lin Y-s, Yin Y-s (2004) A novel one-step chemical method for preparation of copper nanofluids. J Colloid Interface Sci 277(1):100–103
87. Nikkam N et al (2014) Experimental investigation on thermo-physical properties of copper/diethylene glycol nanofluids fabricated via microwave-assisted route. Appl Therm Eng 65(1):158–165

88. Singh AK, Raykar VS (2008) Microwave synthesis of silver nanofluids with polyvinylpyrrolidone (PVP) and their transport properties. Colloid Polym Sci 286(14-15):1667–1673
89. Habibzadeh S et al (2010) Stability and thermal conductivity of nanofluids of tin dioxide synthesized via microwave-induced combustion route. Chem Eng J 156(2):471–478
90. Jalal R et al (2010) ZnO nanofluids: green synthesis, characterization, and antibacterial activity. Mater Chem Phys 121(1–2):198–201
91. Hsu Y-C, Wang W-P, Teng T-P (2016) Fabrication and characterization of nanocarbon-based nanofluids by using an oxygen–acetylene flame synthesis system. Nanoscale Res Lett 11(1):288
92. Kim D et al (2008) Production and characterization of carbon nano colloid via one-step electrochemical method. J Nanopart Res 10(7):1121–1128
93. Kang Z et al (2003) One-step water-assisted synthesis of high-quality carbon nanotubes directly from graphite. J Am Chem Soc 125(45):13652–13653
94. Allen E, Smith P (2001) A review of particle agglomeration. Surfaces 85(86):87
95. Sokolov SV et al (2015) Reversible or not? Distinguishing agglomeration and aggregation at the nanoscale. Anal Chem 87(19):10033–10039
96. Xuan Y, Li Q (2000) Heat transfer enhancement of nanofluids. Int J Heat Fluid Flow 21(1):58–64
97. Li X, Zhu D, Wang X (2007) Evaluation on dispersion behavior of the aqueous copper nano-suspensions. J Colloid Interface Sci 310(2):456–463
98. Kole M, Dey T (2013) Thermal performance of screen mesh wick heat pipes using water-based copper nanofluids. Appl Therm Eng 50(1):763–770
99. Gan Y, Qiao L (2011) Combustion characteristics of fuel droplets with addition of nano and micron-sized aluminum particles. Combust Flame 158(2):354–368
100. Li CH, Peterson G (2006) Experimental investigation of temperature and volume fraction variations on the effective thermal conductivity of nanoparticle suspensions (nanofluids). J Appl Phys 99(8):084314
101. Karthikeyan N, Philip J, Raj B (2008) Effect of clustering on the thermal conductivity of nanofluids. Mater Chem Phys 109(1):50–55
102. Rashin MN, Hemalatha J (2013) Viscosity studies on novel copper oxide–coconut oil nanofluid. Exp Thermal Fluid Sci 48:67–72
103. Kole M, Dey T (2011) Effect of aggregation on the viscosity of copper oxide–gear oil nanofluids. Int J Therm Sci 50(9):1741–1747
104. Subramaniyan A et al (2016) Solar absorption capacity of zinc oxide nanofluids. Curr Sci 111(10):1664–1668
105. Raykar VS, Singh AK (2010) Thermal and rheological behavior of acetylacetone stabilized ZnO nanofluids. Thermochim Acta 502(1):60–65
106. Suganthi KS, Rajan KS (2012) Temperature induced changes in ZnO–water nanofluid: zeta potential, size distribution and viscosity profiles. Int J Heat Mass Transf 55(25–26):7969–7980
107. Saliani M, Jalal R, Goharshadi EK (2015) Effects of pH and temperature on antibacterial activity of zinc oxide nanofluid against *Escherichia coli* O157: H7 and *Staphylococcus aureus*. Jundishapur J Microbiol 8(2):e17115
108. Asadzadeh F, Nasr Esfahany M, Etesami N (2012) Natural convective heat transfer of Fe_3O_4/ethylene glycol nanofluid in electric field. Int J Therm Sci 62:114–119
109. Sheikhbahai M, Nasr Esfahany M, Etesami N (2012) Experimental investigation of pool boiling of Fe_3O_4/ethylene glycol–water nanofluid in electric field. Int J Therm Sci 62:149–153
110. Sundar LS et al (2012) Experimental investigation of forced convection heat transfer and friction factor in a tube with Fe_3O_4 magnetic nanofluid. Exp Thermal Fluid Sci 37:65–71
111. Župan J, Renjo MM (2015) Thermal and rheological properties of water-based ferrofluids and their applicability as quenching media. Phys Procedia 75:1458–1467
112. Manjula S et al (2005) A sedimentation study to optimize the dispersion of alumina nanoparticles in water. Cerâmica 51(318):121–127

113. Nasser MS, James AE (2006) Settling and sediment bed behaviour of kaolinite in aqueous media. Sep Purif Technol 51(1):10–17
114. Yang X-F, Liu Z-H (2011) Pool boiling heat transfer of functionalized nanofluid under sub-atmospheric pressures. Int J Therm Sci 50(12):2402–2412
115. Mehrali M et al (2015) Heat transfer and entropy generation for laminar forced convection flow of graphene nanoplatelets nanofluids in a horizontal tube. Int Commun Heat Mass Transfer 66:23–31
116. Fang Y et al (2016) Synthesis and thermo-physical properties of deep eutectic solvent-based graphene nanofluids. Nanotechnology 27(7):075702
117. Hunter RJ (2013) Zeta potential in colloid science: principles and applications, vol 2. Academic, New York
118. Bose A (2016) Zeta potential for measurement of stability of nanoparticles. Emerging trends of nanotechnology in pharmacy. http://www.pharmainfo.net/book/emerging-trends-nanotechnology-pharmacy-physicochemical-characterization-nanoparticles/zeta. Accessed 7 Feb 2016
119. International Organization for Standardization (2012) ISO 13099: Colloidal systems—methods for zeta potential determination
120. Sahooli M, Sabbaghi S, Shariaty Niassar M (2012) Preparation of CuO/water nanofluids using polyvinylpyrolidone and a survey on its stability and thermal conductivity. Int J Nanosci Nanotechnol 8(1):27–34
121. Jung J-Y, Kim ES, Kang YT (2012) Stabilizer effect on CHF and boiling heat transfer coefficient of alumina/water nanofluids. Int J Heat Mass Transf 55(7):1941–1946
122. Zhang Z et al (2016) Rice husk ash-derived silica nanofluids: synthesis and stability study. Nanoscale Res Lett 11(1):502
123. Haiss W et al (2007) Determination of size and concentration of gold nanoparticles from UV—Vis spectra. Anal Chem 79(11):4215–4221
124. Huang J et al (2009) Influence of pH on the stability characteristics of nanofluids. In: Symposium on photonics and optoelectronics, SOPO 2009, IEEE
125. Xian-Ju W et al (2011) Stability of TiO_2 and Al_2O_3 nanofluids. Chin Phys Lett 28(8):086601
126. Sadeghi R et al (2015) Investigation of alumina nanofluid stability by UV–Vis spectrum. Microfluid Nanofluid 18(5-6):1023–1030
127. Segets D et al (2009) Analysis of optical absorbance spectra for the determination of ZnO nanoparticle size distribution, solubility, and surface energy. ACS Nano 3(7):1703–1710
128. Hwang Y-j et al (2007) Stability and thermal conductivity characteristics of nanofluids. Thermochim Acta 455(1):70–74
129. Subramaniyan A et al (2014) Investigations on the absorption spectrum of TiO_2 nanofluid. J Energy Southern Africa 25(4):123–127
130. Jiang L, Gao L, Sun J (2003) Production of aqueous colloidal dispersions of carbon nanotubes. J Colloid Interface Sci 260(1):89–94
131. Kreibig U, Genzel L (1985) Optical absorption of small metallic particles. Surf Sci 156:678–700
132. Taurozzi JS, Hackley VA, Wiesner M (2012) Preparation of nanoparticle dispersions from powdered material using ultrasonic disruption. NIST Special Publication 1200(2)
133. Hwang Y et al (2008) Production and dispersion stability of nanoparticles in nanofluids. Powder Technol 186(2):145–153
134. Fedele L et al (2011) Experimental stability analysis of different water-based nanofluids. Nanoscale Res Lett 6(1):300
135. Attwood D (2012) Surfactant systems: their chemistry, pharmacy and biology. Springer Science & Business Media, Berlin
136. Yu W, Xie H (2012) A review on nanofluids: preparation, stability mechanisms, and applications. J Nanomater 2012:1
137. Schramm LL (2006) Emulsions, foams, and suspensions: fundamentals and applications. Wiley, New York

138. Parametthanuwat T et al (2011) Application of silver nanofluid containing oleic acid surfactant in a thermosyphon economizer. Nanoscale Res Lett 6(1):315
139. Li X et al (2008) Thermal conductivity enhancement dependent pH and chemical surfactant for cu-H_2O nanofluids. Thermochim Acta 469(1):98–103
140. Wang X-j, Zhu D-s (2009) Investigation of pH and SDBS on enhancement of thermal conductivity in nanofluids. Chem Phys Lett 470(1):107–111
141. Timofeeva EV et al (2010) Particle size and interfacial effects on thermo-physical and heat transfer characteristics of water-based α-SiC nanofluids. Nanotechnology 21(21):215703
142. Yang X, Liu Z-h (2010) A kind of nanofluid consisting of surface-functionalized nanoparticles. Nanoscale Res Lett 5(8):1324
143. Rahmam S, Mohamed N, Sufian S (2014) Effect of acid treatment on the multiwalled carbon nanotubes. Mater Res Innov 18(sup6):S6-196–S6-199
144. Hordy N, Coulombe S, Meunier JL (2013) Plasma functionalization of carbon nanotubes for the synthesis of stable aqueous nanofluids and poly (vinyl alcohol) nanocomposites. Plasma Process Polym 10(2):110–118
145. Tavares J, Coulombe S (2011) Dual plasma synthesis and characterization of a stable copper–ethylene glycol nanofluid. Powder Technol 210(2):132–142
146. US Department of Energy (2011) High thermal conductivity nanofluids offer productivity gains and energy consumption reductions in development and demonstration of nanofluids for industrial cooling applications. DOE/EE-0532
147. Saterlie M et al (2011) Particle size effects in the thermal conductivity enhancement of copper-based nanofluids. Nanoscale Res Lett 6(1)
148. Lu L, Lv L-C, Liu Z-H (2011) Application of cu-water and cu-ethanol nanofluids in a small flat capillary pumped loop. Thermochim Acta 512(1):98–104
149. Tran QH, Le A-T (2013) Silver nanoparticles: synthesis, properties, toxicology, applications and perspectives. Adv Nat Sci Nanosci Nanotechnol 4(3):033001
150. Patel HE et al (2003) Thermal conductivities of naked and monolayer protected metal nanoparticle based nanofluids: manifestation of anomalous enhancement and chemical effects. Appl Phys Lett 83(14):2931–2933
151. Chen H-J, Wen D (2011) Ultrasonic-aided fabrication of gold nanofluids. Nanoscale Res Lett 6(1):198
152. Zhang X, Gu H, Fujii M (2007) Effective thermal conductivity and thermal diffusivity of nanofluids containing spherical and cylindrical nanoparticles. Exp Thermal Fluid Sci 31(6):593–599
153. Shalkevich N et al (2009) On the thermal conductivity of gold nanoparticle colloids. Langmuir 26(2):663–670
154. López-Muñoz GA et al (2012) Thermal diffusivity measurement of spherical gold nanofluids of different sizes/concentrations. Nanoscale Res Lett 7(1):423
155. Kang HU, Kim SH, Oh JM (2006) Estimation of thermal conductivity of nanofluid using experimental effective particle volume. Exp Heat Transfer 19(3):181–191
156. Oliveira GA, Bandarra Filho EP, Wen D (2014) Synthesis and characterization of silver/water nanofluids. High Temp High Pressures 43:69–83
157. Warrier P, Teja A (2011) Effect of particle size on the thermal conductivity of nanofluids containing metallic nanoparticles. Nanoscale Res Lett 6(1):247
158. Parametthanuwat T et al (2015) Experimental investigation on thermal properties of silver nanofluids. Int J Heat Fluid Flow 56:80–90
159. Baladi A, Mamoory RS (2010) Investigation of different liquid media and ablation times on pulsed laser ablation synthesis of aluminum nanoparticles. Appl Surf Sci 256(24):7559–7564
160. Boopathy J et al (2012) Preparation of nano fluids by mechanical method. In: AIP Conference Proceedings, AIP
161. Teipel U, Förter-Barth U (2001) Rheology of nano-scale aluminum suspensions. Propellants Explos Pyrotech 26(6):268–272

162. Barnaud F, Schmelzle P, Schulz P (2000) AQUAZOLE™: an original emulsified water-diesel fuel for heavy-duty applications, SAE Technical Paper
163. Kao M-J et al (2007) Aqueous aluminum nanofluid combustion in diesel fuel. J Test Eval 36(2):1–5
164. Xiu-Tian-Feng E et al (2016) Al-nanoparticle-containing nanofluid fuel: synthesis, stability, properties, and propulsion performance. Ind Eng Chem Res 55(10):2738–2745
165. Huber DL (2005) Synthesis, properties, and applications of iron nanoparticles. Small 1(5):482–501
166. Bashtovoi V, Berkovski B, Bashtovoi V (1996) Magnetic fluids and applications handbook, Series of learning materials. Begell House, New York
167. Hong T-K, Yang H-S, Choi C (2005) Study of the enhanced thermal conductivity of Fe nanofluids. J Appl Phys 97(6):064311
168. Sinha K et al (2009) A comparative study of thermal behavior of iron and copper nanofluids. J Appl Phys 106(6):064307
169. Li Q, Xuan Y, Wang J (2005) Experimental investigations on transport properties of magnetic fluids. Exp Thermal Fluid Sci 30(2):109–116
170. Gan Y, Lim YS, Qiao L (2012) Combustion of nanofluid fuels with the addition of boron and iron particles at dilute and dense concentrations. Combust Flame 159(4):1732–1740
171. Sharifi I, Shokrollahi H, Amiri S (2012) Ferrite-based magnetic nanofluids used in hyperthermia applications. J Magn Magn Mater 324(6):903–915
172. Naphon P, Assadamongkol P, Borirak T (2008) Experimental investigation of titanium nanofluids on the heat pipe thermal efficiency. Int Commun Heat Mass Transfer 35(10):1316–1319
173. Chopkar M et al (2008) Effect of particle size on thermal conductivity of nanofluid. Metall Mater Trans A 39(7):1535–1542
174. Huminic G, Huminic A (2012) Application of nanofluids in heat exchangers: a review. Renew Sust Energ Rev 16(8):5625–5638
175. Saidur R, Leong K, Mohammad H (2011) A review on applications and challenges of nanofluids. Renew Sust Energ Rev 15(3):1646–1668
176. Sui R, Charpentier P (2012) Synthesis of metal oxide nanostructures by direct sol–gel chemistry in supercritical fluids. Chem Rev 112(6):3057–3082
177. Qu Y, Duan X (2013) Progress, challenge and perspective of heterogeneous photocatalysts. Chem Soc Rev 42(7):2568–2580
178. Cheon J, Lee J-H (2008) Synergistically integrated nanoparticles as multimodal probes for nanobiotechnology. Acc Chem Res 41(12):1630–1640
179. Frey NA et al (2009) Magnetic nanoparticles: synthesis, functionalization, and applications in bioimaging and magnetic energy storage. Chem Soc Rev 38(9):2532–2542
180. Kubacka A, Fernández-García M, Colón G (2011) Advanced nanoarchitectures for solar photocatalytic applications. Chem Rev 112(3):1555–1614
181. Tranquada J et al (1995) Evidence for stripe correlations of spins and holes in copper oxide superconductors. Nature 375(6532):561
182. Rout L, Jammi S, Punniyamurthy T (2007) Novel CuO nanoparticle catalyzed C–N cross coupling of amines with iodobenzene. Org Lett 9(17):3397–3399
183. Li S, Eastman J (1999) Measuring thermal conductivity of fluids containing oxide nanoparticles. J Heat Transf 121(2):280–289
184. Kwak K, Kim C (2005) Viscosity and thermal conductivity of copper oxide nanofluid dispersed in ethylene glycol. Kor Aust Rheol J 17(2):35–40
185. Namburu PK et al (2007) Viscosity of copper oxide nanoparticles dispersed in ethylene glycol and water mixture. Exp Thermal Fluid Sci 32(2):397–402
186. Kulkarni DP, Das DK, Chukwu GA (2006) Temperature dependent rheological property of copper oxide nanoparticles suspension (nanofluid). J Nanosci Nanotechnol 6(4):1150–1154
187. Sridhara V, Satapathy LN (2011) Al2O3-based nanofluids: a review. Nanoscale Res Lett 6(1):456

188. Beck MP, Sun T, Teja AS (2007) The thermal conductivity of alumina nanoparticles dispersed in ethylene glycol. Fluid Phase Equilib 260(2):275–278
189. Timofeeva EV et al (2007) Thermal conductivity and particle agglomeration in alumina nanofluids: experiment and theory. Phys Rev E 76(6):061203
190. Esmaeilzadeh E et al (2013) Experimental investigation of hydrodynamics and heat transfer characteristics of γ-Al_2O_3/water under laminar flow inside a horizontal tube. Int J Therm Sci 63:31–37
191. Shankar KS et al (2016) Thermal performance of anodised two phase closed Thermosyphon (TPCT) using Aluminium oxide (Al_2O_3) as nanofluid. Int J ChemTech Res 9(4):239–247
192. Sharma K, Sundar LS, Sarma P (2009) Estimation of heat transfer coefficient and friction factor in the transition flow with low volume concentration of Al_2O_3 nanofluid flowing in a circular tube and with twisted tape insert. Int Commun Heat Mass Transfer 36(5):503–507
193. Teng T-P et al (2010) Thermal efficiency of heat pipe with alumina nanofluid. J Alloys Compd 504:S380–S384
194. Ho C et al (2010) Natural convection heat transfer of alumina-water nanofluid in vertical square enclosures: an experimental study. Int J Therm Sci 49(8):1345–1353
195. Jacob R, Basak T, Das SK (2012) Experimental and numerical study on microwave heating of nanofluids. Int J Therm Sci 59:45–57
196. Carp O, Huisman CL, Reller A (2004) Photoinduced reactivity of titanium dioxide. Prog Solid State Chem 32(1):33–177
197. Barbé CJ et al (1997) Nanocrystalline titanium oxide electrodes for photovoltaic applications. J Am Ceram Soc 80(12):3157–3171
198. Kay A, Grätzel M (1996) Low cost photovoltaic modules based on dye sensitized nanocrystalline titanium dioxide and carbon powder. Sol Energy Mater Sol Cells 44(1):99–117
199. Chen X, Mao SS (2007) Titanium dioxide nanomaterials: synthesis, properties, modifications, and applications. Chem Rev 107(7):2891–2959
200. Brown M, Galley E (1990) Testing UVA and UVB protection from microfine titanium dioxide. Cosmet Toiletries 105(12):69–73
201. Wen D, Ding Y (2005) Formulation of nanofluids for natural convective heat transfer applications. Int J Heat Fluid Flow 26(6):855–864
202. Murshed S, Leong K, Yang C (2005) Enhanced thermal conductivity of TiO_2—water based nanofluids. Int J Therm Sci 44(4):367–373
203. Turgut A et al (2009) Thermal conductivity and viscosity measurements of water-based TiO_2 nanofluids. Int J Thermophys 30(4):1213–1226
204. Du Y-F et al (2010) Breakdown properties of transformer oil-based TiO_2 nanofluid. In: 2010 Annual report conference on electrical insulation and dielectric phenomena (CEIDP), IEEE
205. He Y et al (2007) Heat transfer and flow behaviour of aqueous suspensions of TiO_2 nanoparticles (nanofluids) flowing upward through a vertical pipe. Int J Heat Mass Transf 50(11):2272–2281
206. Chakraborty S et al (2010) Application of water based-TiO_2 nano-fluid for cooling of hot steel plate. ISIJ Int 50(1):124–127
207. Fedele L, Colla L, Bobbo S (2012) Viscosity and thermal conductivity measurements of water-based nanofluids containing titanium oxide nanoparticles. Int J Refrig 35(5):1359–1366
208. Said Z et al (2015) Performance enhancement of a flat plate solar collector using titanium dioxide nanofluid and polyethylene glycol dispersant. J Clean Prod 92:343–353
209. Muthusamy Y et al (2016) Wear analysis when machining AISI 304 with ethylene glycol/ TIO_2 nanoparticle-based coolant. Int J Adv Manuf Technol 82(1-4):327–340
210. Vafaee M, Ghamsari MS (2007) Preparation and characterization of ZnO nanoparticles by a novel sol–gel route. Mater Lett 61(14):3265–3268
211. Yadav RS, Mishra P, Pandey AC (2008) Growth mechanism and optical property of ZnO nanoparticles synthesized by sonochemical method. Ultrason Sonochem 15(5):863–868
212. Seow Z et al (2008) Controlled synthesis and application of ZnO nanoparticles, nanorods and nanospheres in dye-sensitized solar cells. Nanotechnology 20(4):045604

213. Zhang L et al (2008) ZnO nanofluids–a potential antibacterial agent. Prog Nat Sci 18(8):939–944
214. Yu W et al (2009) Investigation of thermal conductivity and viscosity of ethylene glycol based ZnO nanofluid. Thermochim Acta 491(1):92–96
215. Sagadevan S, Shanmugam S (2015) A study of preparation, structural, optical, and thermal conductivity properties of zinc oxide nanofluids. J Nanomed Nanotechnol S6:1
216. Esfe MH, Saedodin S (2014) An experimental investigation and new correlation of viscosity of ZnO–EG nanofluid at various temperatures and different solid volume fractions. Exp Thermal Fluid Sci 55:1–5
217. Pankhurst Q, Jones S, Dobson J (2016) Applications of magnetic nanoparticles in biomedicine: the story so far. J Phys D Appl Phys 49
218. Phuoc TX, Massoudi M (2009) Experimental observations of the effects of shear rates and particle concentration on the viscosity of Fe_2O_3–deionized water nanofluids. Int J Therm Sci 48(7):1294–1301
219. Goshayeshi HR et al (2016) Particle size and type effects on heat transfer enhancement of Ferro-nanofluids in a pulsating heat pipe. Powder Technol 301:1218–1226
220. Salari E et al (2017) Thermal behavior of aqueous iron oxide nano-fluid as a coolant on a flat disc heater under the pool boiling condition. Heat Mass Transf 53(1):265–275
221. Li Z et al (2012) Mesoporous silica nanoparticles in biomedical applications. Chem Soc Rev 41(7):2590–2605
222. Fazeli SA et al (2012) Experimental and numerical investigation of heat transfer in a miniature heat sink utilizing silica nanofluid. Superlattice Microst 51(2):247–264
223. Jin H et al (2014) Properties of mineral oil based silica nanofluids. IEEE Trans Dielectr Electr Insul 21(3):1100–1108
224. Rafati M, Hamidi A, Niaser MS (2012) Application of nanofluids in computer cooling systems (heat transfer performance of nanofluids). Appl Therm Eng 45:9–14
225. Noghrehabadi A, Hajidavaloo E, Moravej M (2016) An experimental investigation on the performance of a symmetric conical solar collector using SiO_2/water nanofluid. Transp Phenom Nano Micro Scales 5(1):23–29
226. Sharif M et al (2017) Preparation and stability of silicone dioxide dispersed in polyalkylene glycol based nanolubricants. In: MATEC Web of Conferences, EDP Sciences
227. Liu Z-h, Liao L (2008) Sorption and agglutination phenomenon of nanofluids on a plain heating surface during pool boiling. Int J Heat Mass Transf 51(9):2593–2602
228. Bagwe RP, Hilliard LR, Tan W (2006) Surface modification of silica nanoparticles to reduce aggregation and nonspecific binding. Langmuir 22(9):4357–4362
229. Xie H, Chen L (2011) Review on the preparation and thermal performances of carbon nanotube contained nanofluids. J Chem Eng Data 56(4):1030–1041
230. Wen D, Ding Y (2004) Effective thermal conductivity of aqueous suspensions of carbon nanotubes (carbon nanotube nanofluids). J Thermophys Heat Transf 18(4):481–485
231. Wusiman K et al (2013) Thermal performance of multi-walled carbon nanotubes (MWCNTs) in aqueous suspensions with surfactants SDBS and SDS. Int Commun Heat Mass Transfer 41:28–33
232. Rashmi W et al (2011) Stability and thermal conductivity enhancement of carbon nanotube nanofluid using gum arabic. J Exp Nanosci 6(6):567–579
233. O'Connell MJ et al (2001) Reversible water-solubilization of single-walled carbon nanotubes by polymer wrapping. Chem Phys Lett 342(3):265–271
234. Assael MJ et al (2005) Thermal conductivity enhancement in aqueous suspensions of carbon multi-walled and double-walled nanotubes in the presence of two different dispersants. Int J Thermophys 26(3):647–664
235. Phuoc TX, Massoudi M, Chen R-H (2011) Viscosity and thermal conductivity of nanofluids containing multi-walled carbon nanotubes stabilized by chitosan. Int J Therm Sci 50(1):12–18
236. Chen L, Xie H (2010) Properties of carbon nanotube nanofluids stabilized by cationic gemini surfactant. Thermochim Acta 506(1–2):62–66

237. Liu J et al (1998) Fullerene pipes. Science 280(5367):1253–1256
238. Zhang L et al (2009) One-step preparation of water-soluble single-walled carbon nanotubes. Appl Surf Sci 255(15):7095–7099
239. Nasiri A et al (2011) Effect of dispersion method on thermal conductivity and stability of nanofluid. Exp Thermal Fluid Sci 35(4):717–723
240. Shin D, Banerjee D (2011) Enhancement of specific heat capacity of high-temperature silica-nanofluids synthesized in alkali chloride salt eutectics for solar thermal-energy storage applications. Int J Heat Mass Transf 54(5):1064–1070
241. Jo B, Banerjee D (2014) Enhanced specific heat capacity of molten salt-based nanomaterials: effects of nanoparticle dispersion and solvent material. Acta Mater 75:80–91
242. Tiznobaik H, Shin D (2013) Enhanced specific heat capacity of high-temperature molten salt-based nanofluids. Int J Heat Mass Transf 57(2):542–548
243. Lu M-C, Huang C-H (2013) Specific heat capacity of molten salt-based alumina nanofluid. Nanoscale Res Lett 8(1):292
244. Chieruzzi M et al (2013) Effect of nanoparticles on heat capacity of nanofluids based on molten salts as PCM for thermal energy storage. Nanoscale Res Lett 8(1):448
245. Ho MX, Pan C (2014) Optimal concentration of alumina nanoparticles in molten Hitec salt to maximize its specific heat capacity. Int J Heat Mass Transf 70:174–184
246. Qiao G et al (2017) Simulation and experimental study of the specific heat capacity of molten salt based nanofluids. Appl Therm Eng 111:1517–1522
247. Bi S-s, Shi L, Zhang L-l (2008) Application of nanoparticles in domestic refrigerators. Appl Therm Eng 28(14):1834–1843
248. Jwo C-S et al (2009) Effects of nanolubricant on performance of hydrocarbon refrigerant system. J Vac Sci Technol B Microelectron Nanometer Struct Process Meas Phenom 27(3):1473–1477
249. Subramani N, Prakash M (2011) Experimental studies on a vapour compression system using nanorefrigerants. Int J Eng Sci Technol 3(9):95–102
250. Anderson LS (2013) Effective thermal conductivity of carbon nanotube-based cryogenic nanofluids. M.S. Thesis, Utah State University

Erratum to: Graphene Analogous Elemental van der Waals Structures

Oswaldo Sanchez, Joung Min Kim, and Ganesh Balasubramanian

Erratum to:
Chapter 4 in: G. Balasubramanian (ed.), *Advances in Nanomaterials*, DOI 10.1007/978-3-319-64717-3_4

Oswaldo Sanchez and Joung Min Kim affiliations were incorrect. The correct information is given below:

Department of Mechanical Engineering, Iowa State University,
Black Engineering Building, Ames, IA 50011-2030, USA

The updated online version of this chapter can be found at
https://doi.org/10.1007/978-3-319-64717-3_4

O. Sanchez (✉) • J.M. Kim
Department of Mechanical Engineering, Iowa State University,
Black Engineering Building, Ames, IA 50011-2030, USA
e-mail: osanchez@iastate.edu

G. Balasubramanian
Lehigh University, Bethlehem, Pennsylvania, USA

G. Balasubramanian (ed.), *Advances in Nanomaterials*,
DOI 10.1007/978-3-319-64717-3_7

Index

G. Balasubramanian (ed.), *Advances in Nanomaterials*,
DOI 10.1007/978-3-319-64717-3

Printed in the United States
By Bookmasters